Introduction to Spectroscopy

Introduction to Spectroscopy

Sheldon Davis

NY RESEARCH
P R E S S

New York

Published by NY Research Press
118-35 Queens Blvd., Suite 400,
Forest Hills, NY 11375, USA
www.nyresearchpress.com

Introduction to Spectroscopy
Sheldon Davis

International Standard Book Number: 978-1-63238-814-8 (Hardback)

Cataloging-in-Publication Data

Introduction to spectroscopy / Sheldon Davis.
    p. cm.
Includes bibliographical references and index.
ISBN 978-1-63238-814-8
1. Spectrum analysis. 2. Spectroscope. 3. Chemistry, Analytic--Qualitative.
I. Davis, Sheldon.
QD95 .I58 2020
543.5--dc23

Contents

Permissions

Index

Preface

The study of interaction between matter and electromagnetic radiation is known as spectroscopy. It measures the radiation intensity as a function of wavelength. Spectroscopy is used as a basic exploratory tool in various fields such as physics, chemistry and astronomy. It allows the electronic structure, composition and physical structure of matter to be analyzed at atomic, macro and molecular scale. Some of the measurement devices that are used in this field are spectrometers, spectrophotometers and spectral analyzers. There are various types of spectroscopy which are characterized by the nature of interaction between the material and energy. A few of these types are absorption spectroscopy, emission spectroscopy, impedance spectroscopy and reflection spectroscopy. The topics included in this book on spectroscopy are of utmost significance and bound to provide incredible insights to readers. Different approaches, evaluations and methodologies related to this field have been included herein. The book is appropriate for students seeking detailed information in this area as well as for experts.

To facilitate a deeper understanding of the contents of this book a short introduction of every chapter is written below:

Chapter 1- The branch of study which focuses on the interaction between electromagnetic radiation and matter is known as spectroscopy. Some of the techniques used in spectroscopy are NMR spectroscopy, Raman Spectroscopy and X-ray Spectroscopy. All these diverse techniques of spectroscopy have been carefully analyzed in this chapter.

Chapter 2- Rotational spectroscopy deals with the measurement of the energies of transitions between quantized rotational states of molecules which are in the gas phase. Its most important application is the exploration of the chemical composition of the interstellar medium through radio telescopes. This chapter closely examines the key concepts of rotational spectrometry such as rotation of polyatomic molecules and microwave rotational spectroscopy.

Chapter 3- Vibrational spectroscopy is a method which is used to measure the vibrational energy of a compound. It is considered to be a non-destructive method of identification. The topics elaborated in this chapter will help in gaining a better perspective about the different aspects of vibrational spectroscopy such as isotope effects.

Chapter 4- Electron spectroscopy deals with the measurement of the kinetic energies which are emitted by electrons when they are bombarded with UV or X-ray radiation. It is used to determine the energy with which electrons are bound in chemical species. This chapter has been carefully written to provide an easy understanding of the varied techniques used within electron spectroscopy such as Auger Electron spectroscopy and electron energy loss spectroscopy.

Chapter 5- Laser spectroscopy is used for the selective excitation of atomic or molecular species. Some of the techniques used within this field are laser Raman spectroscopy, IR laser spectroscopy, ultrafast laser spectroscopy and laser absorption spectroscopy. This chapter discusses in detail these techniques related to laser spectroscopy.

Finally, I would like to thank the entire team involved in the inception of this book for their valuable time and contribution. This book would not have been possible without their efforts. I would also like to thank my friends and family for their constant support.

Sheldon Davis

Understanding Spectroscopy

The branch of study which focuses on the interaction between electromagnetic radiation and matter is known as spectroscopy. Some of the techniques used in spectroscopy are NMR spectroscopy, Raman Spectroscopy and X-ray Spectroscopy. All these diverse techniques of spectroscopy have been carefully analyzed in this chapter.

Spectroscopy is study of the absorption and emission of light and other radiation by matter, as related to the dependence of these processes on the wavelength of the radiation. More recently, the definition has been expanded to include the study of the interactions between particles such as electrons, protons, and ions, as well as their interaction with other particles as a function of their collision energy. Spectroscopic analysis has been crucial in the development of the most fundamental theories in physics, including quantum mechanics, the special and general theories of relativity, and quantum electrodynamics. Spectroscopy, as applied to high-energy collisions, has been a key tool in developing scientific understanding not only of the electromagnetic force but also of the strong and weak nuclear forces.

Spectroscopic techniques have been applied in virtually all technical fields of science and technology. Radio-frequency spectroscopy of nuclei in a magnetic field has been employed in a medical technique called magnetic resonance imaging (MRI) to visualize the internal soft tissue of the body with unprecedented resolution. Microwave spectroscopy was used to discover the so-called three-degree blackbody radiation, the remnant of the big bang (i.e., the primeval explosion) from which the universe is thought to have originated. The internal structure of the proton and neutron and the state of the early universe up to the first thousandth of a second of its existence are being unraveled with spectroscopic techniques using high-energy particle accelerators. The constituentsof distant stars, intergalactic molecules, and even the primordial abundance of the elements before the formation of the first stars can be determined by optical, radio, and X-ray spectroscopy. Optical spectroscopy is used routinely to identify the chemical composition of matter and to determine its physical structure.

Spectroscopic techniques are extremely sensitive. Single atoms and even different isotopes of the same atom can be detected among 10^{20} or more atoms of a different species. (Isotopes are all atoms of an element that have unequal mass but the same atomic number. Isotopes of the same element are virtually identical chemically). Trace amounts of pollutants or contaminants are often detected most effectively by spectroscopic techniques. Certain types of microwave, optical, and gamma-ray spectroscopy are capable of measuring infinitesimal frequency shifts in narrow spectroscopic lines. Frequency shifts as small as one part in 10^{15} of the frequency being measured can be observed with ultrahigh resolution lasertechniques. Because of this sensitivity, the most accurate physical measurements have been frequency measurements.

Spectroscopy now covers a sizable fraction of the electromagnetic spectrum. The table summarizes the electromagnetic spectrum over a frequency range of 16 orders of magnitude. Spectroscopic techniques are not confined to electromagnetic radiation, however. Because the energy E of a photon (a quantum of light) is related to its frequency v by the relation $E = hv$, where h is Planck's constant, spectroscopy is actually the measure of the interaction of photons with matter as a function of the photon energy. In instances where the probe particleis not a photon, spectroscopy refers to the measurement of how the particle interacts with the test particle or material as a function of the energy of the probe particle.

Electromagnetic Phenomena

	Approximate wavelength range (metres)	Approximate frequency range (hertz)
Radio waves	10–1,000	3×10^5–3×10^7
Television waves	1–10	3×10^7–3×10^8
Microwaves, radar	1×10^{-3}–1	3×10^8–3×10^{11}
Infrared	8×10^{-7}–1×10^{-3}	3×10^{11}–4×10^{14}
Visible light	4×10^{-7}–7×10^{-7}	4×10^{14}–7×10^{14}
Ultraviolet	1×10^{-8}–4×10^{-7}	7×10^{14}–3×10^{16}
X-rays	5×10^{-12}–1×10^{-8}	3×10^{16}–6×10^{19}
Gamma rays (γ rays)	$<5 \times 10^{-12}$	$<6 \times 10^{19}$

An example of particle spectroscopy is a surface analysis technique known as electron energy loss spectroscopy (EELS) that measures the energy lost when low-energy electrons (typically 5–10 electron volts) collide with a surface. Occasionally, the colliding electron loses energy by exciting the surface; by measuring the electron's energy loss, vibrational excitations associated with the surface can be measured. On the other end of the energy spectrum, if an electron collides with another particle at exceedingly high energies, a wealth of subatomic particles is produced. Most of what is known in particle physics (the study of subatomic particles) has been gained by analyzing the total particle production or the production of certain particles as a function of the incident energies of electrons and protons.

Optical Spectroscopy

Optical spectroscopy is a means of studying the properties of physical objects based on measuring how an object emits and interacts with light. It can be used to measure attributes such as an object's chemical composition, temperature, and velocity. It involves visible, ultraviolet, or infrared light, alone or in combination, and is part of a larger group of spectroscopic techniques called electromagnetic spectroscopy. Optical spectroscopy is an important technique in modern scientific fields such as chemistry and astronomy.

An object becomes visible by emitting or reflecting photons, and the wavelengths of these photons depend on the object's composition, along with other attributes such as temperature. The human eye perceives the presence and absence of different wavelengths as different colors. For example, photons with a wavelength of 620 to 750 nanometers are perceived as red, and so an object that primarily emits or reflects photons in that range looks red. Using a device called a spectrometer, light can be analyzed with much greater precision. This precise measurement—combined with an understanding of the different properties of light that different substances produce, reflect, or absorb under various conditions—is the basis of optical spectroscopy.

Different chemical elements and compounds vary in how they emit or interact with photons due to quantum mechanical differences in the atoms and molecules that compose them. The light measured by a spectrometer after the light has been reflected from, passed through, or emitted by the object being studied has what are called spectral lines. These lines are sharp discontinuities of light or darkness in the spectrum that indicate unusually high or unusually low numbers of photons of particular wavelengths. Different substances produce distinctive spectral lines that can be used to identify them. These spectral lines are also affected by factors such as the object's temperature and velocity, so spectroscopy can also be used to measure these as well. In addition to wavelength, other characteristics of the light, such as its intensity, can also provide useful information.

Optical spectroscopy can be done in several different ways, depending on what is being studied. Individual spectrometers are specialized devices that focus on precise analysis of specific, narrow parts of the electromagnetic spectrum. They therefore exist in a wide variety of types for different applications.

One major type of optical spectroscopy, called absorption spectroscopy, is based on identifying which wavelengths of light a substance absorbs by measuring the photons it allows to pass through. The light can be produced specifically for this purpose with equipment such as lamps or lasers or may come from a natural source, such as starlight. It is most commonly used with gases, which are diffuse enough to interact with light while still allowing it to pass through. Absorption spectroscopy is useful for identifying chemicals and can be used to differentiate elements or compounds in a mixture.

This method is also extremely important in modern astronomy and is often used to study the temperature and chemical composition of celestial objects. Astronomical spectroscopy also measures the velocity of distant objects by taking advantage of the Doppler effect. Light waves from an object that is moving toward the observer appear to have higher frequencies and thus lower wavelengths than light waves from an object at rest relative to the observer, while the waves from an object that is moving away appear to have lower frequencies. These phenomena are called blueshift and redshift, respectively, because raising the frequency of a wave of visible light moves it toward the blue/violet end of the spectrum, while lowering the frequency moves it toward red.

Another important form of optical spectroscopy is called emissions spectroscopy. When atoms or molecules are excited by an outside energy source such as light or heat, they temporarily increase in energy level before dropping back to their ground state. When the excited particles return to their ground state, they release the excess energy in the form of photons. As is the case with absorption, different substances emit photons of different wavelengths that can then be measured and analyzed. In one common form of this technique, called fluorescence spectroscopy, the subject being analyzed is energized with light, usually ultraviolet light. In atomic emissions spectroscopy, fire, electricity, or plasma is used.

Fluorescence spectroscopy is commonly used in biology and medicine, as it is less damaging to biological materials than other methods and because some organic molecules are naturally fluorescent. Atomic absorption spectroscopy is used in chemical analysis and is particularly effective for detecting metals. Different types of atomic absorption spectroscopy are used for purposes such as identifying valuable minerals in ores, analyzing evidence from crime scenes, and maintaining quality control in metallurgy and industry.

NMR Spectroscopy

Nuclear Magnetic Resonance (NMR) spectroscopy is an analytical chemistry technique used in quality control and reserach for determining the content and purity of a sample as well as itsmolecular structure. For example, NMR can quantitatively analyze mixtures containing known compounds. For unknown compounds, NMR can either be used to match against spectral libraries or to infer the basic structure directly. Once the basic structure is known, NMR can be used to determine molecular conformation in solution as well as studying physical properties at the molecular level such as conformational exchange, phase changes, solubility, and diffusion. In order to achieve the desired results, a variety of NMR techniques are available.

The basis of NMR

The principle behind NMR is that many nuclei have spin and all nuclei are electrically charged. If an external magnetic field is applied, an energy transfer is possible between the base energy to a higher energy level (generally a single energy gap). The energy transfer takes place at a wavelength that corresponds to radio frequencies and when the spin returns to its base level, energy is emitted at the same frequency. The signal that matches this transfer is measured in many ways and processed in order to yield an NMR spectrum for the nucleus concerned.

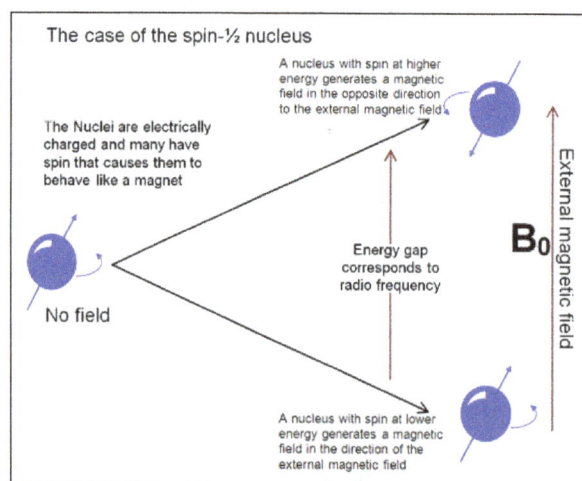

The basis of NMR.

In the above figure, relates to spin-½ nuclei that include the most commonly used NMR nucleus, proton (^{1}H or hydrogen-1) as well as many other nuclei such as ^{13}C, ^{15}N and ^{31}P. Many nuclei such as deuterium (^{2}H or hydrogen-2) have a higher spin and are therefore quadrupolar and although they yield NMR spectra their energy diagram and some of their properties are different.

Chemical Shift

The precise resonant frequency of the energy transition is dependent on the effective magnetic field at the nucleus. This field is affected by electron shielding which is in turn dependent on the chemical environment. As a result, information about the nucleus' chemical environment can be derived from its resonant frequency. In general, the more electronegative the nucleus is, the higher the resonant frequency. Other factors such as ring currents (anisotropy) and bond strain affect the frequency shift. It is customary to adopt tetramethylsilane (TMS) as the proton reference frequency. This is because the precise resonant frequency shift of each nucleus depends on the magnetic field used. The frequency is not easy to remember (for example, the frequency of benzene might be 400.132869 MHz) so it was decided to define chemical shift as follows to yield a more convenient number such as 7.17 ppm:

$$\delta = (\nu - \nu_0)/\nu_0$$

The chemical shift, using this equation, is not dependent on the magnetic field and it is convenient to express it in ppm where (for proton) TMS is set to ν_0 thereby giving it a chemical shift of zero. For other nuclei, ν_0 is defined as $\varXi \, \nu_{TMS}$ where $\varXi$ is the frequency ratio of the nucleus (e. g., 25.145020% for ^{13}C).

In the case of the ^{1}H NMR spectrum of ethyl benzene, the methyl (CH_3) group is the most electron withdrawing (electronegative) and therefore resonates at the lowest chemical shift. The aromatic phenyl group is the most electron donating (electropositive) so has the highest chemical shift. The methylene (CH_2) falls somewhere in the middle. However, if the chemical shift of the aromatics were due to electropositivity alone, then they would resonate between four and five ppm. The increased chemical shift is due to the delocalized ring current of the phenyl group.

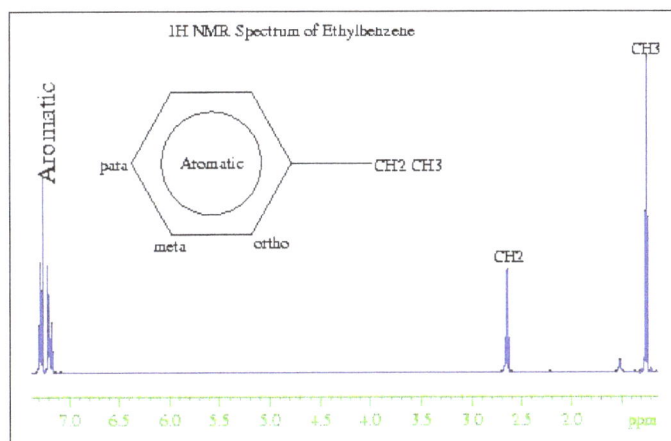

Figure: ^{1}H NMR spectrum of ethylbenznene.

This definition of chemical shift is sufficient for most purposes. However, complications arise when comparing chemical shifts under different conditions: solvent, temperature, *etc.* Chemical shifts are affected slightly by isotopic substitution, an effect that is known as an isotope shift.

Spin-spin Coupling

The effective magnetic field is also affected by the orientation of neighboring nuclei. This effect is known as spin-spin coupling which can cause splitting of the signal for each type of nucleus into two or more lines.

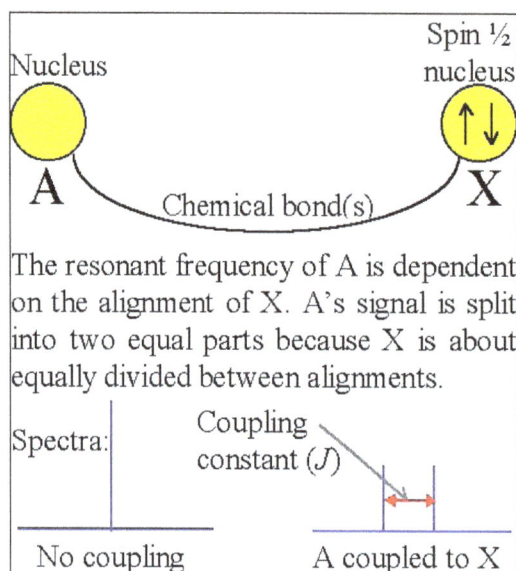

Figure: Spin-spin coupling.

The size of the splitting (coupling constant or J) is independent of the magnetic field and is therefore measured as an absolute frequency (usually Hertz). The number of splittings indicates the number of chemically bonded nuclei in the vicinity of the observed nucleus. Some common coupling patterns are shown below.

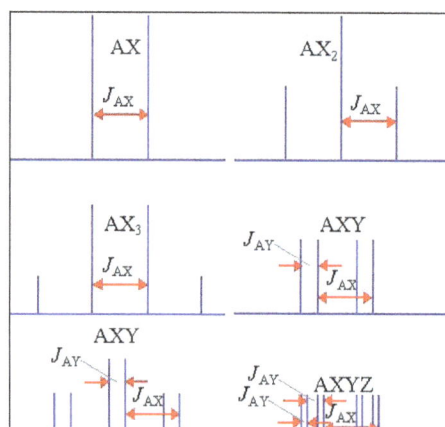

Figure: Examples of coupling patterns showing coupling constants.

The above patterns are a first order approximation and are correct provided that all the coupled spins have widely separated chemical shifts. The different nuclei are labeled with the letters A and X (in a system of this type the letters come from widely separated parts of the alphabet). If the chemical shifts are similar then distortions in peak height occur as in the diagram below (the letters are also close together in the alphabet). For more than two spins, extra signals may appear. These effects are called second order coupling.

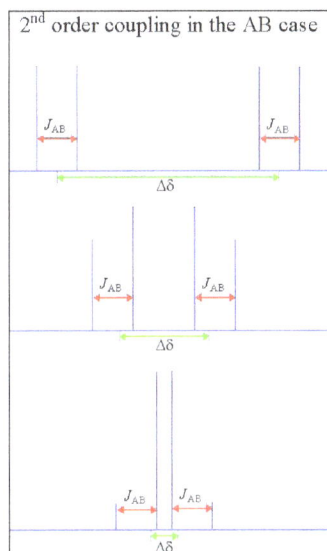

Figure: An example of second order coupling.

Returning to the example of ethylbenzene, the methyl (CH_3) group has a coupling pattern in the form of A_3X_2 which to a first order approximation looks like an AX_2 multiplet. Likewise, the methylene (CH_2) group has the form A_2X_3 that is equivalent to AX_3. The first order approximation works because the groups are widely separated in the spectrum. The aromatic signals are close together and display second order effects. The *ortho* signal is a doublet AX while the *meta* and *para* signals are triplets.

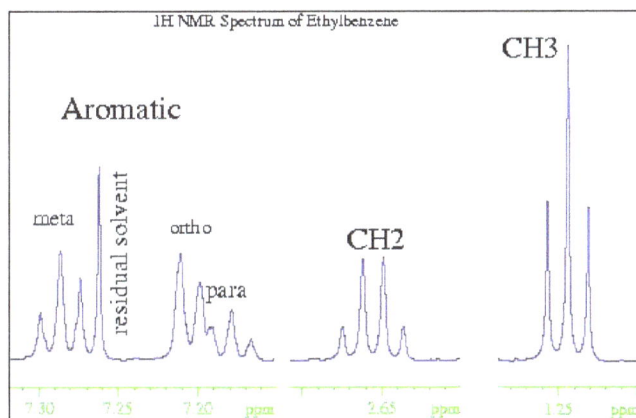

Figure: Couplings in the ethylbenzene spectrum.

Proton NMR Spectroscopy

This important and well-established application of nuclear magnetic resonance will serve to illustrate some of the novel aspects of this method. To begin with, the nmr spectrometer must be tuned to a specific nucleus, in this case the proton. The actual procedure for obtaining the spectrum varies, but the simplest is referred to as the continuous wave (CW) method. A typical CW-spectrometer is shown in the following diagram. A solution of the sample in a uniform 5 mm glass tube is oriented between the poles of a powerful magnet, and is spun to average any magnetic field variations, as well as tube imperfections. Radio frequency radiation of appropriate energy is broadcast into the sample from an antenna coil (colored red). A receiver coil surrounds the sample tube, and emission of absorbed rf energy is monitored by dedicated electronic devices and a computer. An nmr spectrum is acquired by varying or sweeping the magnetic field over a small range while observing the rf signal from the sample. An equally effective technique is to vary the frequency of the rf radiation while holding the external field constant.

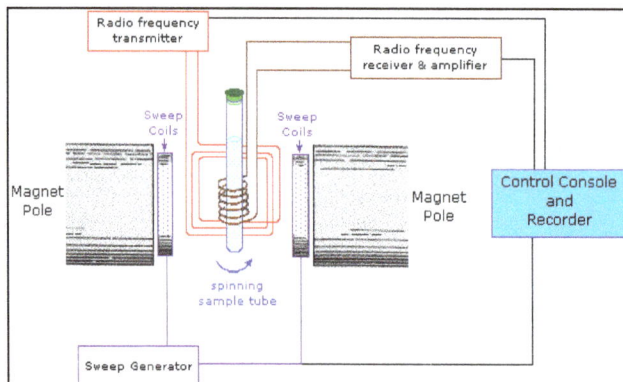

As an example, consider a sample of water in a 2.3487 T external magnetic field, irradiated by 100 MHz radiation. If the magnetic field is smoothly increased to 2.3488 T, the hydrogen nuclei of the water molecules will at some point absorb rf energy and a resonance signal will appear. An animation showing this may be activated by clicking the Show Field Sweep button. The field sweep will be repeated three times, and the resulting resonance trace is colored red. For visibility, the water proton signal displayed in the animation is much broader than it would be in an actual experiment.

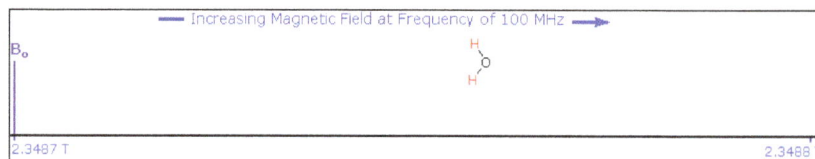

Since protons all have the same magnetic moment, we might expect all hydrogen atoms to give resonance signals at the same field / frequency values. Fortunately for chemistry applications, this is not true. It is not possible, of course, to examine isolated protons in the spectrometer described above; but from independent measurement and calculation it has been determined that a naked proton would resonate at a lower field strength than the nuclei of covalently bonded hydrogens. With the exception of water, chloroform and sulfuric acid, which are examined as liquids, all the other compounds are measured as gases.

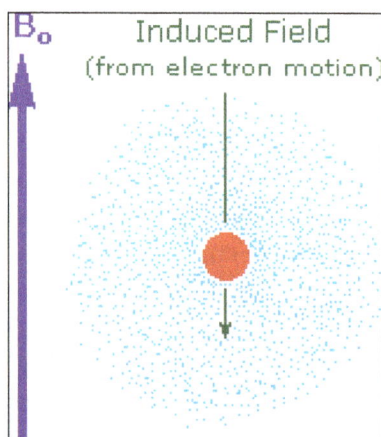

The proton nuclei in different compounds behave differently in the nmr experiment. Since electrons are charged particles, they move in response to the external magnetic field (B_o) so as to generate a secondary field that opposes the much stronger applied field. This secondary field shields the nucleus from the applied field, so B_o must be increased in order to achieve resonance (absorption of rf energy). As illustrated in the drawing on the right, B_o must be increased to compensate for the induced shielding field. In the upper diagram, those compounds that give resonance signals at the higher field side of the diagram (CH_4, HCl, HBr and HI) have proton nuclei that are more shielded than those on the lower field (left) side of the diagram. The magnetic field range displayed in the above diagram is very small compared with the actual field strength (only about 0.0042%). It is customary to refer to small increments such as this in units of parts per million (ppm). The difference between 2.3487 T and 2.3488 T is therefore about 42 ppm. Instead of designating a range of nmr signals in terms of magnetic field differences (as above), it is more common to use a frequency scale, even though the spectrometer may operate by sweeping the magnetic field. Using this terminology, we would find that at 2.34 T the proton signals shown above extend over a 4,200 Hz range (for a 100 MHz rf frequency, 42 ppm is 4,200 Hz). Most organic compounds exhibit proton resonances that fall within a 12 ppm range (the shaded area), and it is therefore necessary to use very sensitive and precise spectrometers to resolve structurally distinct sets of hydrogen atoms within this narrow range. In this respect it might be noted that the detection of a part-per-million difference is equivalent to detecting a 1 millimeter difference in distances of 1 kilometer.

Chemical Shift

Unlike infrared and uv-visible spectroscopy, where absorption peaks are uniquely located by a frequency or wavelength, the location of different nmr resonance signals is dependent on both the external magnetic field strength and the rf frequency. Since no two magnets will have exactly the same field, resonance frequencies will vary accordingly and an alternative method for characterizing and specifying the location of nmr signals is needed. This problem is illustrated by the eleven different compounds shown in the following diagram. Although the eleven resonance signals are distinct and well separated, an unambiguous numerical locator cannot be directly assigned to each.

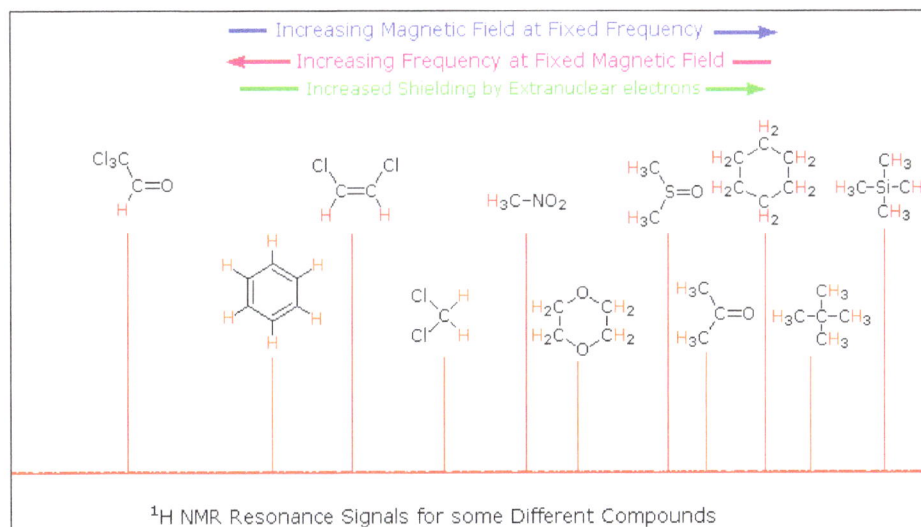

^{1}H NMR Resonance Signals for some Different Compounds

One method of solving this problem is to report the location of an nmr signal in a spectrum relative to a reference signal from a standard compound added to the sample. Such a reference standard should be chemically unreactive, and easily removed from the sample after the measurement. Also, it should give a single sharp nmr signal that does not interfere with the resonances normally observed for organic compounds. Tetramethylsilane, $(CH_3)_4Si$, usually referred to as TMS, meets all these characteristics, and has become the reference compound of choice for proton and carbon nmr. Since the separation (or dispersion) of nmr signals is magnetic field dependent, one additional step must be taken in order to provide an unambiguous location unit. To correct these frequency differences for their field dependence, we divide them by the spectrometer frequency (100 or 500 MHz in the example), The resulting number would be very small, since we are dividing Hz by MHz, so it is multiplied by a million. Note that v_{ref} is the resonant frequency of the reference signal and v_{samp} is the frequency of the sample signal. This operation gives a locator number called the Chemical Shift, having units of parts-per-million (ppm), and designated by the symbol δ.

The compounds referred to above share two common characteristics:

- The hydrogen atoms in a given molecule are all structurally equivalent, averaged for fast conformational equilibria.

- The compounds are all liquids, save for neopentane which boils at 9 °C and is a liquid in an ice bath.

The first feature assures that each compound gives a single sharp resonance signal. The second allows the pure (neat) substance to be poured into a sample tube and examined in a nmr spectrometer. In order to take the nmr spectra of a solid, it is usually necessary to dissolve it in a suitable solvent. Early studies used carbon tetrachloride for this purpose, since it has no hydrogen that could introduce an interfering signal. Unfortunately, CCl_4 is a poor solvent for many polar compounds and is also toxic. Deuterium labeled compounds, such as deuterium oxide (D_2O), chloroform-d ($DCCl_3$), benzene-d_6 (C_6D_6), acetone-d_6 (CD_3COCD_3) and DMSO-d_6 (CD_3SOCD_3) are now widely used as nmr solvents. Since the deuterium isotope of hydrogen has a different magnetic moment and spin, it is invisible in a spectrometer tuned to protons.

From the example given ,it may deduce that one factor contributing to chemical shift differences in proton resonance is the inductive effect. If the electron density about a proton nucleus is relatively high, the induced field due to electron motions will be stronger than if the electron density is relatively low. The shielding effect in such high electron density cases will therefore be larger, and a higher external field (B_0) will be needed for the rf energy to excite the nuclear spin. Since silicon is less electronegative than carbon, the electron density about the methyl hydrogens in tetramethylsilane is expected to be greater than the electron density about the methyl hydrogens in neopentane (2,2-dimethylpropane), and the characteristic resonance signal from the silane derivative does indeed lie at a higher magnetic field. Such nuclei are said to be shielded. Elements that are more electronegative than carbon should exert an opposite effect (reduce the electron density); and, as the data in the following tables show, methyl groups bonded to such elements display lower field signals (they are deshielded). The deshielding effect of electron withdrawing groups is roughly proportional to their electronegativity, as shown by the left table. Furthermore, if more than one such group is present, the deshielding is additive (table on the right), and proton resonance is shifted even further downfield.

Proton Chemical Shifts of Methyl Derivatives					Proton Chemical Shifts (ppm)					
Compound	$(CH_3)_4C$	$(CH_3)_3N$	$(CH_3)_2O$	CH_3F	Cpd./Sub.	X=Cl	X=Br	X=I	X=OR	X=SR
δ	0.9	2.1	3.2	4.1	CH_3X	3.0	2.7	2.1	3.1	2.1
Compound	$(CH_3)_4Si$	$(CH_3)_3P$	$(CH_3)_2S$	CH_3Cl	CH_2X_2	5.3	5.0	3.9	4.4	3.7
δ	0.0	0.9	2.1	3.0	CHX_3	7.3	6.8	4.9	5.0	

The general distribution of proton chemical shifts associated with different functional groups is summarized in the following chart. Bear in mind that these ranges are approximate, and may not encompass all compounds of a given class. Note also that the ranges specified for OH and NH protons (colored orange) are wider than those for most CH protons. This is due to hydrogen bonding variations at different sample concentrations.

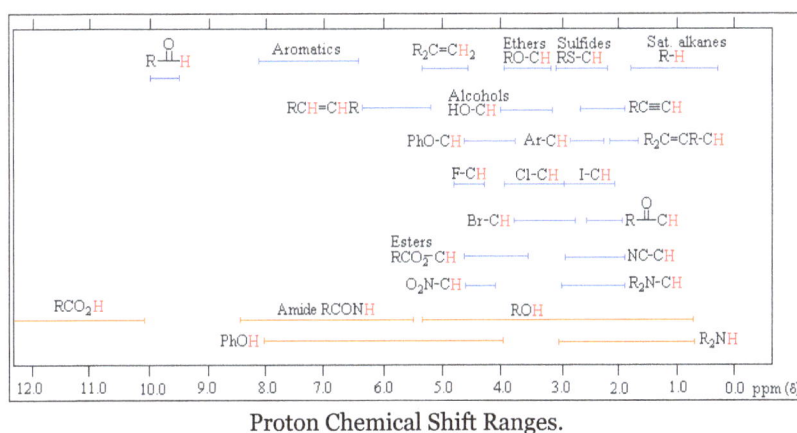

Proton Chemical Shift Ranges.

Signal Strength

The magnitude or intensity of nmr resonance signals is displayed along the vertical axis of a spectrum and is proportional to the molar concentration of the sample. Thus, a small or dilute sample will give a weak signal, and doubling or tripling the sample concentration increases the signal strength proportionally. If we take the nmr spectrum of equal molar amounts of benzene and cyclohexane in carbon tetrachloride solution, the resonance signal from cyclohexane will be twice as intense as that from benzene because cyclohexane has twice as many hydrogens per molecule. This is an important relationship when samples incorporating two or more different sets of hydrogen atoms are examined, since it allows the ratio of hydrogen atoms in each distinct set to be determined. To this end it is necessary to measure the relative strength as well as the chemical shift of the resonance signals that comprise an nmr spectrum. Two common methods of displaying the integrated intensities associated with a spectrum are illustrated by the following examples. In the three spectra in the top row, a horizontal integrator trace (light green) rises as it crosses each signal by a distance proportional to the signal strength. Alternatively, an arbitrary number, selected by the instrument's computer to reflect the signal strength, is printed below each resonance peak, as shown in the three spectra in the lower row. From the relative intensities shown here, together with the previously noted chemical shift correlations, the reader should be able to assign the signals in these spectra to the set of hydrogens that generates each. When evaluating relative signal strengths, it is useful to set the smallest integration to unity and convert the other values proportionally.

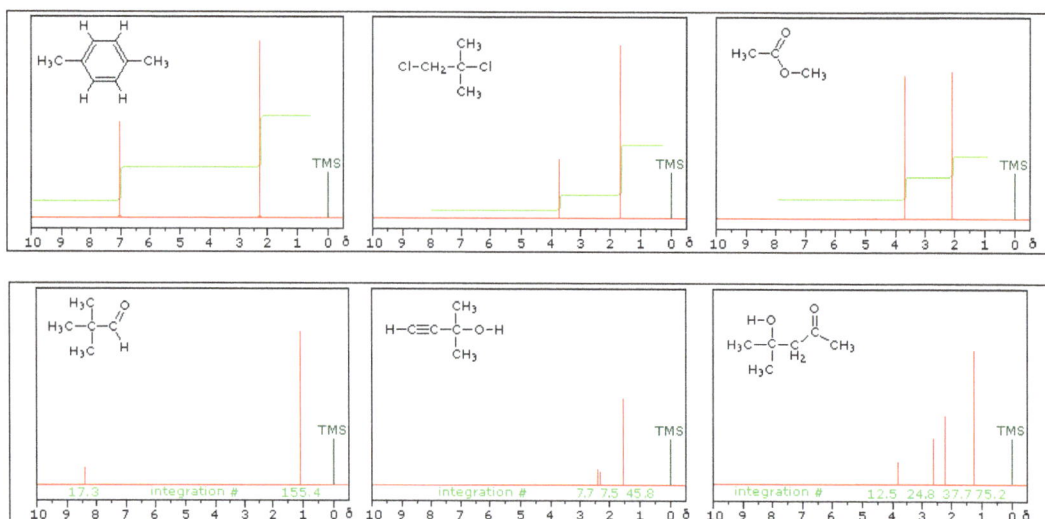

Hydroxyl Proton Exchange and the Influence of Hydrogen Bonding

The last two compounds in the lower row are alcohols. The OH proton signal is seen at 2.37 δ in 2-methyl-3-butyne-2-ol, and at 3.87 δ in 4-hydroxy-4-methyl-2-pentanone, illustrating the wide range over which this chemical shift may be found. A six-membered ring intramolecular hydrogen bond in the latter compound is in part responsible for its low field shift. We can take advantage of rapid OH exchange with the deuterium of heavy water to assign hydroxyl proton resonance signals. As shown in the following equation, this removes the hydroxyl proton from the sample and its resonance signal in the nmr spectrum disappears. Experimentally, one simply adds a drop of heavy water to a chloroform-d solution of the compound and runs the spectrum again. The result of this exchange is displayed below:

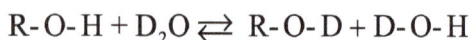

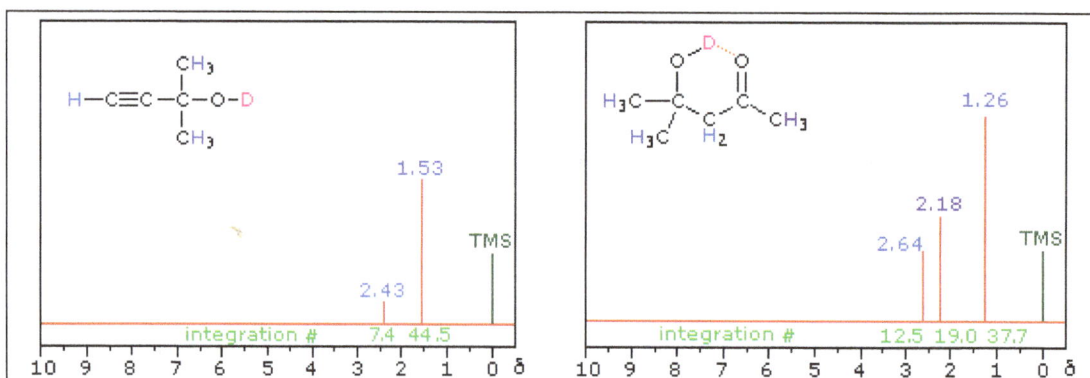

$$R\text{-}O\text{-}H + D_2O \rightleftharpoons R\text{-}O\text{-}D + D\text{-}O\text{-}H$$

Hydrogen bonding shifts the resonance signal of a proton to lower field (higher frequency). Numerous experimental observations support this statement.

1. The chemical shift of the hydroxyl hydrogen of an alcohol varies with concentration. Very dilute solutions of 2-methyl-2-propanol, $(CH_3)_3COH$, in carbon tetrachloride solution display a hydroxyl resonance signal having a relatively high-field chemical shift (< 1.0 δ). In concentrated solution this signal shifts to a lower field, usually near 2.5 δ.

2. The more acidic hydroxyl group of phenol generates a lower-field resonance signal, which shows a similar concentration dependence to that of alcohols. OH resonance signals for different percent concentrations of phenol in chloroform-d are shown in the following diagram (C-H signals are not shown).

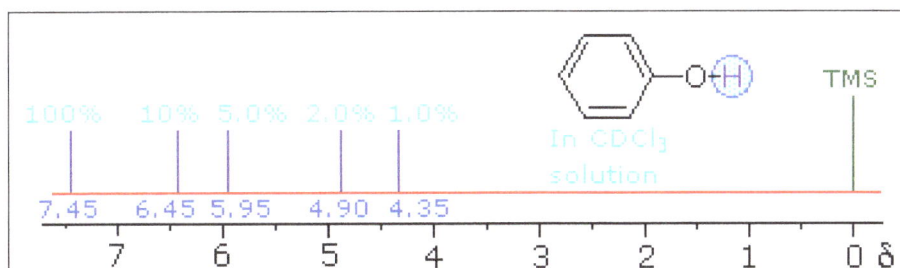

3. Because of their favored hydrogen-bonded dimeric association, the hydroxyl proton of carboxylic acids displays a resonance signal significantly down-field of other functions. For a typical acid it appears from 10.0 to 13.0 δ and is often broader than other signals. The spectra shown below for chloroacetic acid (left) and 3,5-dimethylbenzoic acid (right) are examples.

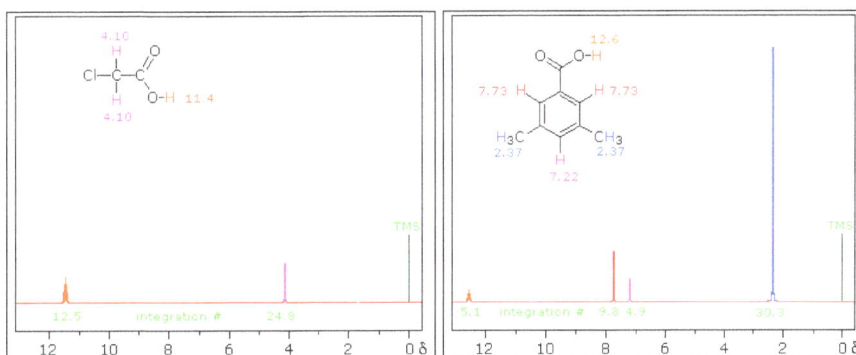

4. Intramolecular hydrogen bonds, especially those defining a six-membered ring, generally display a very low-field proton resonance. The case of 4-hydroxypent-3-ene-2-one (the enol tautomer of 2,4-pentanedione) not only illustrates this characteristic, but also provides an instructive example of the sensitivity of the nmr experiment to dynamic change. In the nmr spectrum of the pure liquid, sharp signals from both the keto and enol tautomers are seen, their mole ratio being 4 : 21 (keto tautomer signals are colored purple). Chemical shift assignments for these signals are shown in the shaded box above the spectrum. The chemical shift of the hydrogen-bonded hydroxyl proton is δ 14.5, exceptionally downfield. We conclude, therefore, that the rate at which these tautomers interconvert is slow compared with the inherent time scale of nmr spectroscopy.

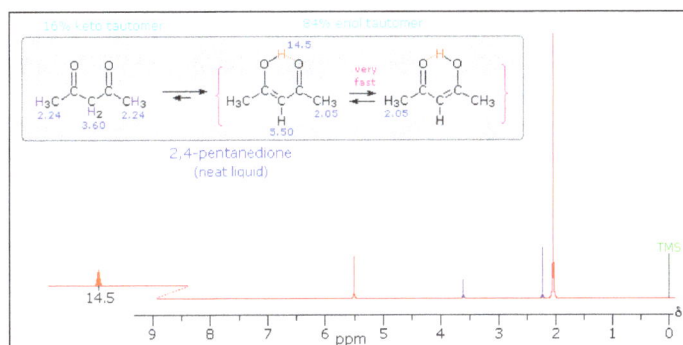

Two structurally equivalent structures may be drawn for the enol tautomer (in magenta brackets). If these enols were slow to interconvert, we would expect to see two methyl resonance signals associated with each, one from the allylic methyl and one from the methyl ketone. Since only one strong methyl signal is observed, we must conclude that the interconversion of the enols is very fast-so fast that the nmr experiment detects only a single time-averaged methyl group (50% α-keto and 50% allyl).

Although hydroxyl protons have been the focus of this discussion, it should be noted that corresponding N-H groups in amines and amides also exhibit hydrogen bonding nmr shifts, although to a lesser degree. Furthermore, OH and NH groups can undergo rapid proton exchange with each other; so if two or more such groups are present in a molecule, the nmr spectrum will show a single signal at an average chemical shift. For example, 2-hydroxy-2-methylpropanoic acid, $(CH_3)_2C(OH)CO_2H$, displays a strong methyl signal at δ 1.5 and a 1/3 weaker and broader OH signal at δ 7.3 ppm. Note that the average of the expected carboxylic acid signal (ca. 12) and the alcohol signal (ca. 2) is 7. Rapid exchange of these hydrogens with heavy water, as noted above, would cause the low field signal to disappear.

π-Electron Functions

An examination of the proton chemical shift chart makes it clear that the inductive effect of substituents cannot account for all the differences in proton signals. In particular the low field resonance of hydrogens bonded to double bond or aromatic ring carbons is puzzling, as is the very low field signal from aldehyde hydrogens. The hydrogen atom of a terminal alkyne, in contrast, appears at a relatively higher field. All these anomalous cases seem to involve hydrogens bonded to pi-electron systems, and an explanation may be found in the way these pi-electrons interact with the applied magnetic field.

Pi-electrons are more polarizable than are sigma-bond electrons, as addition reactions of electrophilic reagents to alkenes testify. Therefore, we should not be surprised to find that field induced pi-electron movement produces strong secondary fields that perturb nearby nuclei. The pi-electrons associated with a benzene ring provide a striking example of this phenomenon. The electron cloud above and below the plane of the ring circulates in reaction to the external field so as to generate an opposing field at the center of the ring and a supporting field at the edge of the ring. This kind of spatial variation is called anisotropy, and it is common to nonspherical distributions of electrons, as are found in all the functions mentioned above. Regions in which the induced field supports or adds to the external field are said to be deshielded, because a slightly weaker external field will bring about resonance for nuclei in such areas. However, regions in which the induced field opposes the external field are termed shielded because an increase in the applied field is needed for resonance. Shielded regions are designated by a plus sign, and deshielded regions by a negative sign.

Note that the anisotropy about the triple bond nicely accounts for the relatively high field chemical shift of ethynyl hydrogens. The shielding & deshielding regions about the carbonyl group have been described in two ways, which alternate in the display.

Sigma bonding electrons also have a less pronounced, but observable, anisotropic influence on nearby nuclei. This is seen in the small deshielding shift that occurs in the series $CH_3 - R, R - CH_2 - R, R_3CH$; as well as the deshielding of equatorial versus axial protons on a fixed cyclohexane ring.

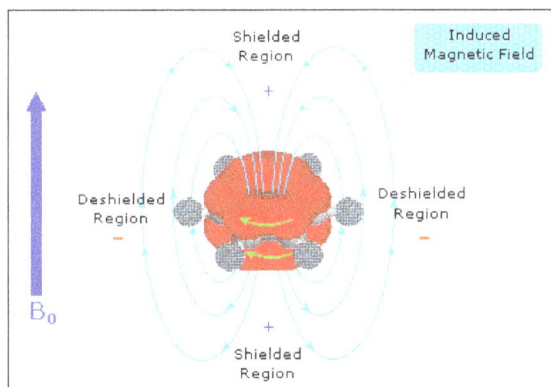

Solvent Effects

Chloroform-d ($CDCl_3$) is the most common solvent for nmr measurements, thanks to its good solubilizing character and relative unreactive nature (except for 1° and 2°-amines). Other deuterium labeled compounds, such as deuterium oxide (D_2O), benzene-d6 (C_6D_6), acetone-d6 (CD_3COCD_3) and DMSO-d6 (CD_3SOCD_3) are also available for use as nmr solvents. Because some of these solvents have π-electron functions and/or may serve as hydrogen bonding partners, the chemical shifts of different groups of protons may change depending on the solvent being used. The following table gives a few examples, obtained with dilute solutions at 300 MHz.

Some Typical ¹H Chemical Shifts (δ values) in Selected Solvents						
Solvent Compound	$CDCl_3$	C_6D_6	CD_3COCD_3	CD_3SOCD_3	$CD_3C{\equiv}N$	D_2O
$(CH_3)_3C{-}O{-}CH_3$ $C{-}CH_3$ $O{-}CH_3$	1.19 3.22	1.07 3.04	1.13 3.13	1.11 3.03	1.14 3.13	1.21 3.22
$(CH_3)_3C{-}O{-}H$ $C{-}CH_3$ $O{-}H$	1.26 1.65	1.05 1.55	1.18 3.10	1.11 4.19	1.16 2.18	--- ---
$C_6H_5CH_3$ CH_3 C_6H_5	2.36 7.15-7.20	2.11 7.00-7.10	2.32 7.10-7.20	2.30 7.10-7.15	2.33 7.15-7.30	--- ---
$(CH_3)_2C{=}O$	2.17	1.55	2.09	2.09	2.08	2.22

For most of the above resonance signals and solvents the changes are minor, being on the order of ±0.1 ppm. However, two cases result in more extreme changes and these have provided useful applications in structure determination. First, spectra taken in benzene-d$_6$ generally show small upfield shifts of most C–H signals, but in the case of acetone this shift is about five times larger than normal. Carbonyl groups form weak π–π collision complexes with benzene rings, that persist long enough to exert a significant shielding influence on nearby groups. In the case of substituted cyclohexanones, axial α-methyl groups are shifted upfield by 0.2 to 0.3 ppm; whereas equatorial methyls are slightly deshielded (shift downfield by about 0.05 ppm). These changes are all relative to the corresponding chloroform spectra. The second noteworthy change is seen in the spectrum of tert-butanol in DMSO, where the hydroxyl proton is shifted 2.5 ppm down-field from where it is found in dilute chloroform solution. This is due to strong hydrogen bonding of the alcohol O–H to the sulfoxide oxygen, which not only de-shields the hydroxyl proton, but secures it from very rapid exchange reactions that prevent the display of spin-spin splitting. Similar but weaker hydrogen bonds are formed to the carbonyl oxygen of acetone and the nitrogen of acetonitrile.

Spin-Spin Interactions

The nmr spectrum of 1,1-dichloroethane is more complicated than we might have expected from the previous examples. Unlike its 1,2-dichloro-isomer, which displays a single resonance signal from the four structurally equivalent hydrogens, the two signals from the different hydrogens are split into close groupings of two or more resonances. This is a common feature in the spectra of compounds having different sets of hydrogen atoms bonded to adjacent carbon atoms. The signal splitting in proton spectra is usually small, ranging from fractions of a Hz to as much as 18 Hz, and is designated as J (referred to as the coupling constant). In the 1,1-dichloroethane example all the coupling constants are 6.0 Hz.

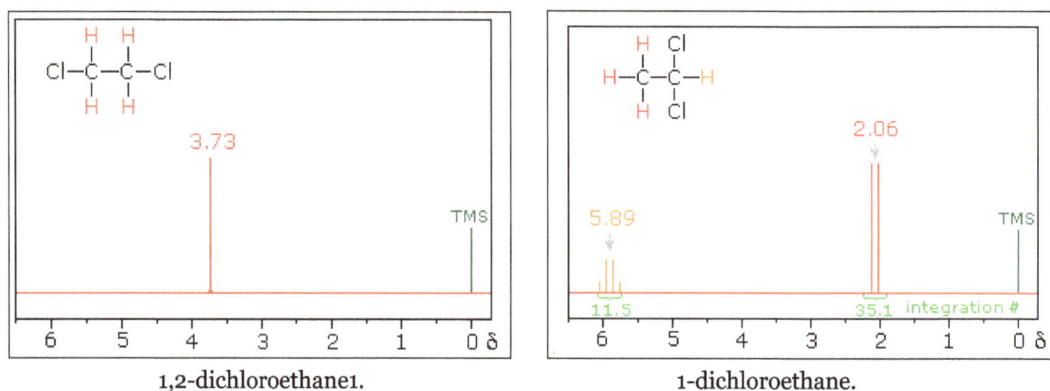

1,2-dichloroethane1.   1-dichloroethane.

The splitting patterns found in various spectra are easily recognized, provided the chemical shifts of the different sets of hydrogen that generate the signals differ by two or more ppm. The patterns are symmetrically distributed on both sides of the proton chemical shift, and the central lines are always stronger than the outer lines. The most commonly observed patterns have been given descriptive names, such as doublet (two equal intensity signals), triplet (three signals with an intensity ratio of 1:2:1) and quartet (a set of four signals with intensities of 1:3:3:1). Four such patterns are displayed in the following illustration. The line separation is always constant within a given multiplet, and is called the coupling constant (J). The magnitude of J, usually given in units of Hz, is magnetic field independent.

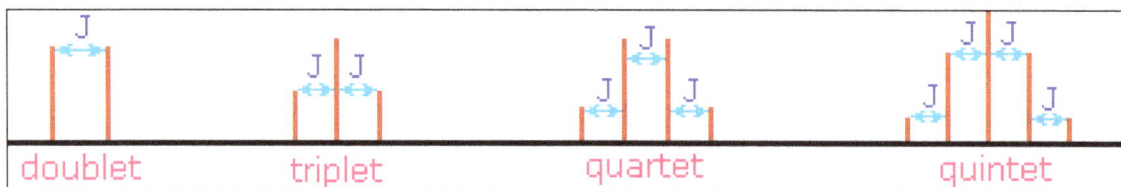

The splitting patterns shown above display the ideal or "First-Order" arrangement of lines. This is usually observed if the spin-coupled nuclei have very different chemical shifts (i.e. Δv is large compared to J). If the coupled nuclei have similar chemical shifts, the splitting patterns are distorted (second order behavior). In fact, signal splitting disappears if the chemical shifts are the same.

Two examples that exhibit minor 2nd order distortion are shown below (both are taken at a frequency of 90 MHz). The ethyl acetate spectrum on the left displays the typical quartet and triplet of a substituted ethyl group. The spectrum of 1,3-dichloropropane on the right demonstrates that equivalent sets of hydrogens may combine their influence on a second, symmetrically located set. Even though the chemical shift difference between the A and B protons in the 1,3-dichloroethane spectrum is fairly large (140 Hz) compared with the coupling constant (6.2 Hz), some distortion of the splitting patterns is evident. The line intensities closest to the chemical shift of the coupled partner are enhanced. Thus the B set triplet lines closest to A are increased, and the A quintet lines nearest B are likewise stronger. A smaller distortion of this kind is visible for the A and C couplings in the ethyl acetate spectrum.

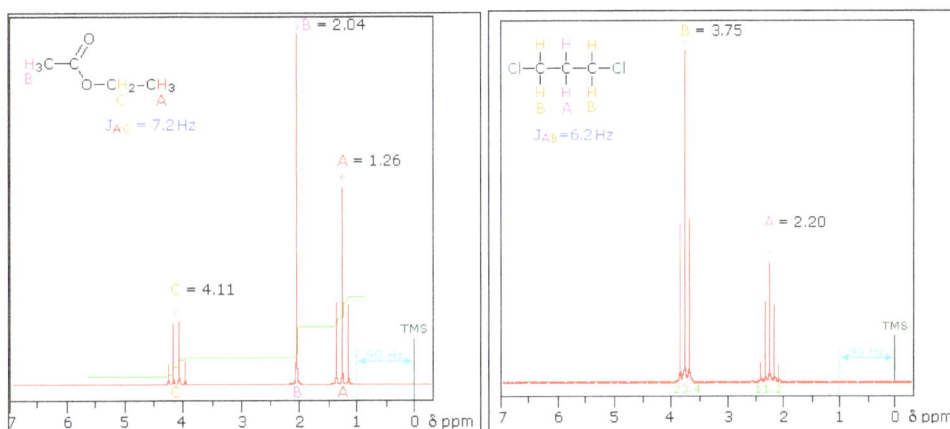

Information Obtained from the Signal Splitting

If an atom under examination is perturbed or influenced by a nearby nuclear spin (or set of spins), the observed nucleus responds to such influences, and its response is manifested in its resonance signal. This spin-coupling is transmitted through the connecting bonds, and it functions in both directions. Thus, when the perturbing nucleus becomes the observed nucleus, it also exhibits signal splitting with the same J. For spin-coupling to be observed, the sets of interacting nuclei must be bonded in relatively close proximity (e.g. vicinal and geminal locations), or be oriented in certain optimal and rigid configurations. Some spectroscopists place a number before the symbol J to designate the number of bonds linking the coupled nuclei (colored orange below). Using this terminology, a vicinal coupling constant is ^3J and a geminal constant is ^2J.

Geminal Hydrogens   Vicinal Hydrogens

The following general rules summarize important requirements and characteristics for spin 1/2 nuclei:

1) Nuclei having the same chemical shift (called isochronous) do not exhibit spin-splitting. They may actually be spin-coupled, but the splitting cannot be observed directly. 2) Nuclei separated

by three or fewer bonds (e.g. vicinal and geminal nuclei) will usually be spin-coupled and will show mutual spin-splitting of the resonance signals (same J's), provided they have different chemical shifts. Longer-range coupling may be observed in molecules having rigid configurations of atoms. 3) The magnitude of the observed spin-splitting depends on many factors and is given by the coupling constant J (units of Hz). J is the same for both partners in a spin-splitting interaction and is independent of the external magnetic field strength. 4) The splitting pattern of a given nucleus (or set of equivalent nuclei) can be predicted by the n+1 rule, where n is the number of neighboring spin-coupled nuclei with the same (or very similar) Js. If there are 2 neighboring, spin-coupled, nuclei the observed signal is a triplet (2+1=3); if there are three spin-coupled neighbors the signal is a quartet (3+1=4). In all cases the central line(s) of the splitting pattern are stronger than those on the periphery. The intensity ratio of these lines is given by the numbers in Pascal's triangle. Thus a doublet has 1:1 or equal intensities, a triplet has an intensity ratio of 1:2:1, a quartet 1:3:3:1 etc. To see how the numbers in Pascal's triangle are related to the Fibonacci series.

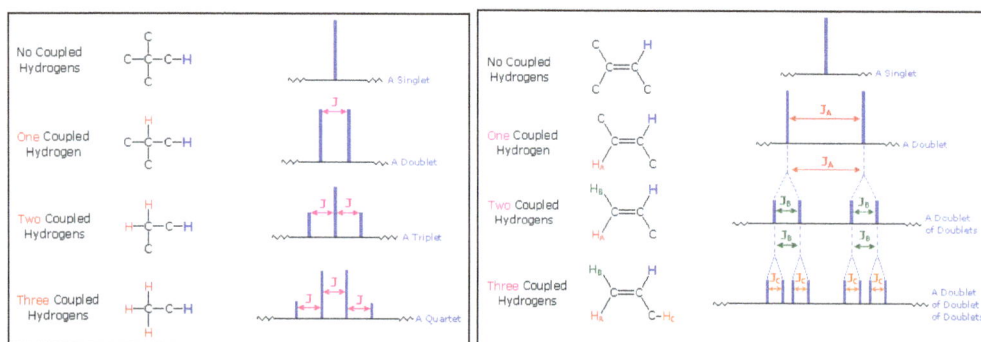

If a given nucleus is spin-coupled to two or more sets of neighboring nuclei by different J values, the n+1 rule does not predict the entire splitting pattern. Instead, the splitting due to one J set is added to that expected from the other J sets. Bear in mind that there may be fortuitous coincidence of some lines if a smaller J is a factor of a larger J.

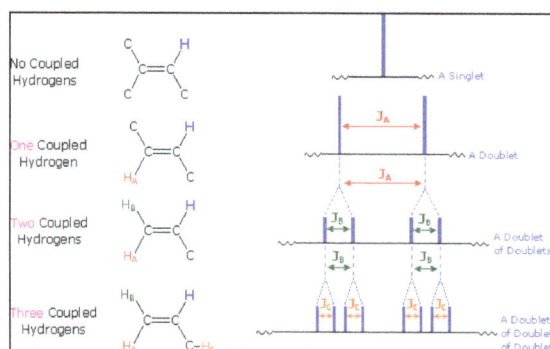

Magnitude of Some Typical Coupling Constants

Spin 1/2 nuclei include ^1H, ^{13}C, ^{19}F & ^{31}P. The spin-coupling interactions described above may occur between similar or dissimilar nuclei. If, for example, a ^{19}F is spin-coupled to a ^1H, both nuclei will appear as doublets having the same J constant. Spin coupling with nuclei having spin other than 1/2 is more complex.

Structural Type	J (Hz)	Structural Type	J (Hz)
$H-\overset{\mid}{C}-(C)_n-\overset{\mid}{C}-H$	0 (unless in a rigid ideal orientation)		12 to 18
H_3C-CH_2-X	6 to 8		7 to 12
	5 to 7		0.5 to 3
	2 to 12 (depends on dihedral angle and the nature of X and Y)		3 to 11 (depends on dihedral angle)
	0.5 to 3	$-\overset{\mid}{C}-C\equiv C-H$	2 to 3
	12 to 15 (must be diastereotopic)		o 6 to 9 m 1 to 3 p 0 to 1

Carbon NMR Spectroscopy

The power and usefulness of ^{1}H nmr spectroscopy as a tool for structural analysis should be evident from the past discussion. Unfortunately, when significant portions of a molecule lack C-H bonds, no information is forthcoming. Examples include polychlorinated compounds such as chlordane, polycarbonyl compounds such as croconic acid, and compounds incorporating triple bonds (structures below, orange colored carbons).

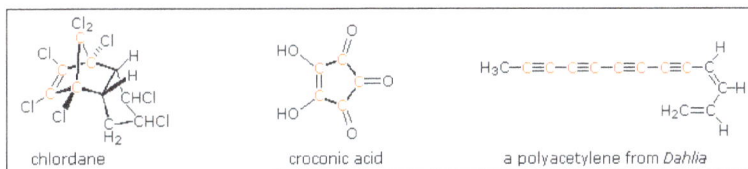

chlordane   croconic acid   a polyacetylene from *Dahlia*

Even when numerous C-H groups are present, an unambiguous interpretation of a proton nmr spectrum may not be possible. The following diagram depicts three pairs of isomers (A & B) which display similar proton nmr spectra. Although a careful determination of chemical shifts should permit the first pair of compounds (blue box) to be distinguished, the second and third cases (red & green boxes) might be difficult to identify by proton nmr alone.

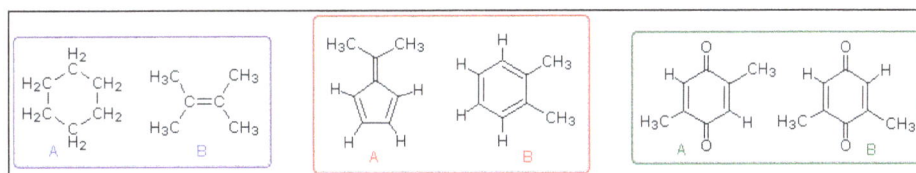

These difficulties would be largely resolved if the carbon atoms of a molecule could be probed by nmr in the same fashion as the hydrogen atoms. Since the major isotope of carbon (^{12}C) has no spin, this option seems unrealistic. Fortunately, 1.1% of elemental carbon is the ^{13}C isotope, which has a spin I = 1/2, so in principle it should be possible to conduct a carbon nmr experiment. It is worth noting here, that if much higher abundances of ^{13}C were naturally present in all carbon compounds, proton nmr would become much more complicated due to large one-bond coupling of ^{13}C and ^{1}H.

The most important operational technique that has led to successful and routine ^{13}C nmr spectroscopy is the use of high-field pulse technology coupled with broad-band heteronuclear decoupling of all protons. The results of repeated pulse sequences are accumulated to provide improved signal strength. Also, for reasons that go beyond the present treatment, the decoupling irradiation enhances the sensitivity of carbon nuclei bonded to hydrogen.

When acquired in this manner, the carbon nmr spectrum of a compound displays a single sharp signal for each structurally distinct carbon atom in a molecule (remember, the proton couplings have been removed). The spectrum of camphor, shown on the left below, is typical. Furthermore, a comparison with the ^{1}H nmr spectrum on the right illustrates some of the advantageous characteristics of carbon nmr. The dispersion of ^{13}C chemical shifts is nearly twenty times greater than that for protons, and this together with the lack of signal splitting makes it more likely that every structurally distinct carbon atom will produce a separate signal. The only clearly identifiable signals in the proton spectrum are those from the methyl groups. The remaining protons have resonance signals between 1.0 and 2.8 ppm from TMS, and they overlap badly thanks to spin-spin splitting.

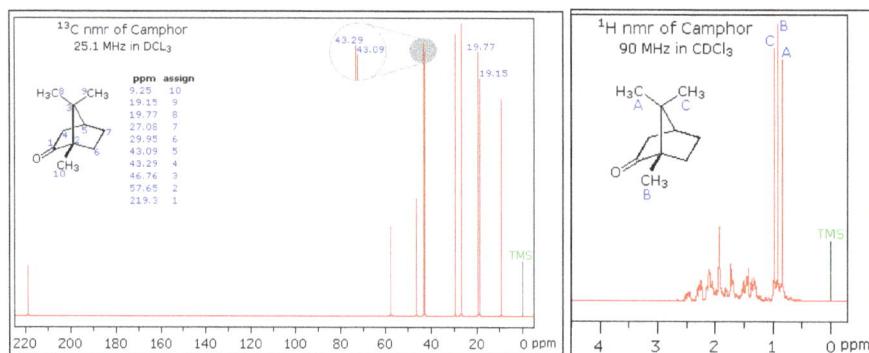

Unlike proton nmr spectroscopy, the relative strength of carbon nmr signals are not normally proportional to the number of atoms generating each one. Because of this, the number of discrete signals and their chemical shifts are the most important pieces of evidence delivered by a carbon spectrum. The general distribution of carbon chemical shifts associated with different functional groups is summarized in the following chart. Bear in mind that these ranges are approximate, and may not encompass all compounds of a given class. Note also that the over 200 ppm range of chemical shifts shown here is much greater than that observed for proton chemical shifts.

^{13}C Chemical Shift Ranges

Low Field Region

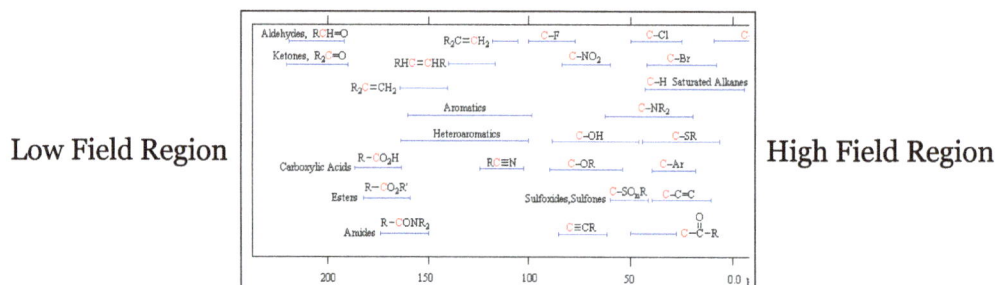

High Field Region

For samples in CDCl$_3$ solution. The δ scale is relative to TMS at δ=0.

The isomeric pairs previously cited as giving very similar proton nmr spectra are now seen to be distinguished by carbon nmr. In the example on the left below (blue box), cyclohexane and 2,3-dimethyl-2-butene both give a single sharp resonance signal in the proton nmr spectrum (the former at δ 1.43 ppm and the latter at 1.64 ppm). However, in its carbon nmr spectrum cyclohexane displays a single signal at δ 27.1 ppm, generated by the equivalent ring carbon atoms (colored blue); whereas the isomeric alkene shows two signals, one at δ 20.4 ppm from the methyl carbons (colored brown), and the other at 123.5 ppm (typical of the green colored sp^2 hybrid carbon atoms).

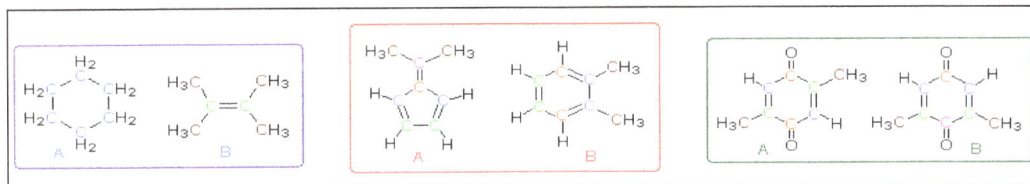

The C_8H_{10} isomers in the center (red) box have pairs of homotopic carbons and hydrogens, so symmetry should simplify their nmr spectra. The fulvene (isomer A) has five structurally different groups of carbon atoms (colored brown, magenta, orange, blue and green respectively) and should display five ^{13}C nmr signals (one near 20 ppm and the other four greater than 100 ppm). Although ortho-xylene (isomer B) will have a proton nmr very similar to isomer A, it should only display four ^{13}C nmr signals, originating from the four different groups of carbon atoms (colored brown, blue, orange and green). The methyl carbon signal will appear at high field (near 20 ppm), and the aromatic ring carbons will all give signals having $\delta > 100$ ppm. Finally, the last isomeric pair, quinones A & B in the green box, are easily distinguished by carbon nmr. Isomer A displays only four carbon nmr signals (δ 15.4, 133.4, 145.8 & 187.9 ppm); whereas, isomer B displays five signals (δ 15.9, 133.3, 145.8, 187.5 & 188.1 ppm), the additional signal coming from the non-identity of the two carbonyl carbon atoms (one colored orange and the other magenta).

Ultraviolet and Visible Spectroscopy

Ultraviolet-visible spectroscopy or ultraviolet-visible spectrophotometry (UV-Vis or UV/Vis) refers to absorption spectroscopy in the ultraviolet-visible spectral region. This means it uses light in the visible and adjacent (near-UV and near-infrared (NIR)) ranges.

Ultraviolet and visible (UV-Vis) absorption spectroscopy is the measurement of the attenuation (weakening of strength) of a beam of light after it passes through a sample or after reflection from a sample surface.

Ultraviolet and visible light are energetic enough to promote outer electrons to higher energy sublevels. The technique is usually applied to molecules or inorganic complexes in solution. Absorption measurements can be at a singlewavelength or over an extended spectral range.

UV-Vis spectra have broad features that are of limited use for sample identification but are very useful for quantitative measurements. The concentration of an analyte in solution can be determined by measuring the absorbance at some wavelength and applying the Beer-Lambert Law.

Since the UV-Vis range spans the range of human visual acuity of approximately 400 - 750 nm, UV-Vis spectroscopy is useful to characterize the absorption, transmission, and reflectivity of a variety of technologically important materials, such as pigments, coatings, windows, and filters. This more qualitative application usually requires recording at least a portion of the UV-Vis spectrum for characterization of the optical or electronic properties of materials.

Instrumentation

1. The UV-Vis spectral range is approximately 190 to 900 nm, as defined by the working range of typical commercial UV-Vis spectrophotometers.

2. The short-wavelength limit for simple UV-Vis spectrometers is the absorption of ultraviolet wavelengths less than 180 nm by atmospheric gases. Purging a spectrometer with nitrogen gas extends this limit to 175 nm. Working beyond 175 nm requires a vacuum spectrometer and a suitable UV light source.

3. The long-wavelength limit is usually determined by the wavelength response of the detector in the spectrometer. High-end commercial UV-Vis spectrophotometers extend the measurable spectral range into the NIR region as far as 3300 nm.

4. The light source is usually a deuterium discharge lamp for UV measurements and a tungsten-halogen lamp for visible and NIR measurements. The instruments automatically swap lamps when scanning between the UV and visible regions. The wavelengths of these continuous light sources are typically dispersed by a holographic grating in a single or double monochromator or spectrograph. The spectral bandpass is then determined by the monochromator slit width or by the array-element width in array-detector spectrometers.

5. Spectrometer designs and optical components are optimized to reject stray light, which is one of the limiting factors in quantitative absorbance measurements.

6. The detector in single-detector instruments is a photodiode, phototube, or photomultiplier tube (PMT). UV-Vis-NIR spectrometers utilize a combination of a PMT and a Peltier-cooled PbS IR detector. The light beam is redirected automatically to the appropriate detector when scanning between the visible and NIR regions. The diffraction grating and instrument parameters such as slit width can also change.

Most commercial UV-Vis absorption spectrometers use one of three overall optical designs: a fixed or scanning spectrometer with a single light beam and sample holder, a scanning spectrometer with dual light beams and dual sample holders for simultaneous measurement of P and Po, or a non-scanning spectrometer with an array detector for simultaneous measurement of multiple wavelengths. In single-beam and dual-beam spectrometers, the light from a lamp is dispersed before reaching the sample cell. In an array-detector instrument, all wavelengths pass through the sample and the dispersing element is between the sample and the array detector.

Raman Spectroscopy

Raman spectroscopy is based on the inelastic light scattering in a substance where the incident light transfers energy to molecular vibrations. The scattered light can be detected by a Raman spectrometer and represents a "chemical fingerprint" of the substance. Based on such spectral information, a material can be identified or characterized.

Properties of Light

Light is a form of electromagnetic radiation, which has both wave and particle ("photon") properties. Light waves are usually mathematically described by a cosine function, where the two most important characteristic parameters are the wavelength (distance between two consecutive wave crests or troughs) and the amplitude (height of the waves over the baseline).

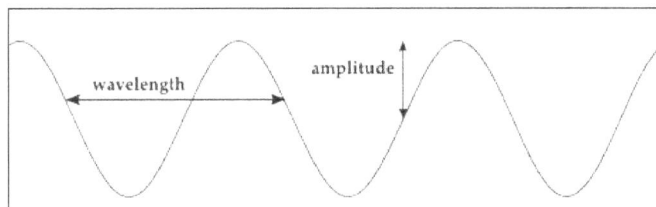

Figure: Light can be described as a wave.

Electromagnetic radiation propagates through space carrying a certain amount of electromagnetic energy. This energy is proportional to its frequency v of oscillation, which is connected to the wavelength λ by the speed of light c.

$$v = \frac{c}{\lambda}$$

where v= frequency, λ = wavelength, c = speed of light.

Therefore, a light wave (or photon) carries more energy E the larger the frequency or, alternatively, the smaller the wavelength is,

$$E \propto v \propto \frac{1}{\lambda}$$

where : v = frequency, λ= wavelength, E = energy.

For historical reasons, spectroscopists also like to use the wavenumber $\tilde{v}$, which is defined as the reciprocal of the wavelength. The wavenumber is directly proportional to the energy of the photon and usually expressed in units of reciprocal centimeters (cm^{-1}) to give easy to read numbers:

$$\tilde{v} = \frac{1}{\lambda} = \frac{v}{c}$$

where v = frequency, λ = wavelength, $\tilde{v}$ = wavenumber, c = speed of light.

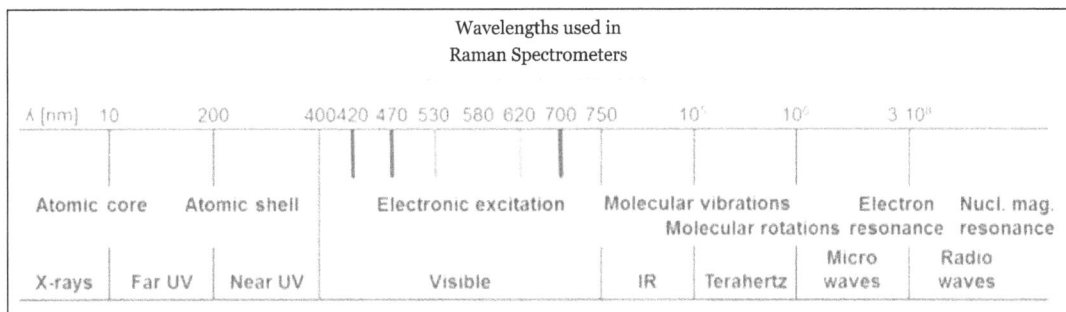

Figure: Electromagnetic spectrum: Depending on the energy of the electromagnetic radiation, different processes in atoms and molecules can be induced by the interaction between light and matter.

The above description is valid for a single light wave or photon. However, a light beam consists of many light waves with different frequencies propagating in the same direction. Each frequency contributes to the beam with intensity I (i.e. a certain number of photons per time interval). The intensity of a light beam is the quantity that is ultimately measured with the detector of a spectrometer.

The intensity distribution of all frequencies is called the spectrum of this light beam. Only a small part of the light frequencies can be seen by the human eye ("visible light"). Other spectral regions are e.g. microwave, infrared, ultra-violet (UV), or Röntgen (X-ray) radiation. For Raman spectroscopy, visible light or infrared (IR) light is used for the excitation.

The most important physical parameters and their corresponding equations relevant for Raman spectroscopy are summarized in table.

Table: Summary of light parameters.

Parameter	Symbol	Relation	Unit	Explanation
Energy	E	$E \propto \nu \propto \dfrac{1}{\lambda}$	J(Joule)	Amount of energy contained in a single light wave
Frequency	ν	$\nu = \dfrac{c}{\lambda}$	HZ = 1/s (Hertz)	Rate of oscillation of the light wave, determines its energy
Wavelength	λ	$\lambda = \dfrac{c}{\nu}$	Nm (Nanometer)	Spatial distance between two wave crests (or troughs)
Wavenumber	$\tilde{\nu}$	$\tilde{\nu} = \dfrac{1}{\lambda} = \dfrac{\nu}{c}$	cm^{-1} (Reciprocal Centimeters)	Energy parmeter commonly used in spectroscopy
Light intensity	I	Not important in this context	Not important in this context	Number of photons of a certain frequency contained in a light beam

Light-matter Interactions

When a light beam hits matter, it will interact with it in a specific way, dependent on the interplay between the light waves and the atoms and molecules that make up the matter. The interaction may leave the energy of matter and light unchanged (e.g. refraction, reflection, elastic scattering) or lead to an energy exchange between both. The processes used in spectroscopy to characterize matter belong to the latter category.

The transfer of energy from light to matter leads to an excitation. important excitation processes required to understand Raman spectroscopy: absorption, fluorescence, and scattering.

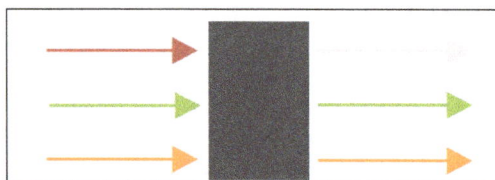

Figure: Absorption: Some of the incident wavelengths are (partially) absorbed in the sample, while other wavelengths are transmitted without much loss in intensity.

- Absorption: Light energy in some parts of the electromagnetic spectrum is (partially) transferred to the matter. This means some light waves pass through the matter without modification (transmission), while some light is absorbed by the sample.

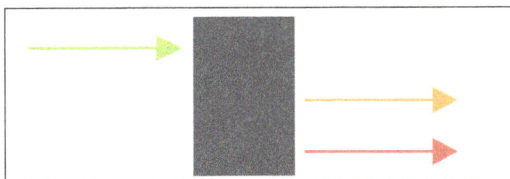

Figure: Fluorescence: The incident green light is absorbed (resulting in a loss in intensity) and reemitted in different, longer wavelengths (this results in a gain in intensity or even addition of new "colors").

- Fluorescence: Matter can reemit absorbed light again by an independent process called fluorescence. The emitted light has a different, longer wavelength than the originally absorbed light, which will result in a perceived "amplification" of some light wavelengths. This is a particularly common process for colored samples and large biomolecules.

Scattering

When an intense light source (e.g. a laser) hits a sample, part of the light will be scattered in different directions. The majority of the scattered light has the same wavelength ("color") as the incident light – it is elastically scattered. This is why the human eye is able to "see" the spot of a laser pointer on the wall or table.

However, a tiny fraction of the scattered light interacts with the matter it hits in a way that it exchanges small amounts of energy, which is called inelastic scattering. The change in energy of the scattered light results in a changed frequency and wavelength.

The microscopic origin of this Raman interaction is an excitation or de-excitation of molecular vibrations in the matter. The characteristics of these vibrations determine the wavelength of the inelastically scattered light. From measuring the intensity distribution (spectrum) of the scattered light it is hence possible to deduce information about the vibrational structure of the substance illuminated. Therefore, Raman spectroscopy belongs to the group of vibrational spectroscopies.

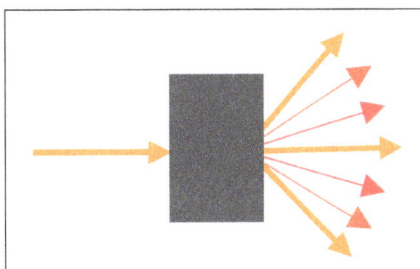

Figure: Raman scattering: Most of the incident yellow light is scattered elastically in all directions. Small amounts of light, usually with higher wavelengths (orange, red), are also scattered inelastically after interaction with the molecules of the sample.

Each of these processes can be exploited to extract information about the chemical and physical nature of the sample. The exact type and extent of molecular properties deducible depends on the type of spectroscopy used. The two main vibrational spectroscopies are infrared (IR) spectroscopy and Raman spectroscopy.

Raman Scattering Theory

There are three scattering processes that are important for Raman spectroscopy:

- Rayleigh scattering is the term used for elastic scattering of light by molecules, and is by far the most dominant scattering process. The interaction does not change the energy state of the molecule and as such the scattered photon has the same color (wavelength) as the incident photon. In a Raman spectrometer, the Rayleigh scattered light has to be removed from the collected light, otherwise it would obscure the Raman signals.

- Stokes Raman scattering is the inelastic scattering process that transfers energy from the light to a vibration of the molecule. Therefore, the scattered photon has lower energy and a higher wavelength than the incident photon. The amount of energy transferred is not arbitrary, it has to be exactly the amount required to excite one of the molecular vibrations of the molecule. The composition of the scattered light is therefore highly dependent on the exact type of molecule (like a fingerprint). Stokes scattering is the most commonly exploited process to acquire a Raman spectrum. It is, however, several orders of magnitude less likely to occur compared to Rayleigh scattering, rendering it difficult to detect.

- Anti-Stokes Raman scattering is another inelastic scattering process. Here, a specific amount of energy is transferred from a molecular vibration to the photon. The scattered photon has higher energy and a lower wavelength than the incident photon. This process is even less likely to occur than Stokes scattering. Therefore, it is usually not used in Raman spectroscopy. The information extracted from anti-Stokes scattered light is mostly equivalent to the information extracted from Stokes scattered light, and only very specialized applications will require the extra effort to measure both scattering processes.

Table: Summarizes the scattering processes relevant for Raman spectroscopy.

	Rayleigh	Stokes (Raman)	Anti-Stokes
Scattering is	elastic	inelastic	inelastic
Energy transfer	none	Photon → Molecule	Molecule → Photon
Effect on molecule	None	Excitation of vibrations or rotations	De-excitation of preexcited vibrations or rotations
Effect on photon	change in direction, same wavelength	change in direction, higher wavelength	change in direction, lower wavelength
Probability of occurrence	common	very rare	extremely rare

Vibrational Spectroscopy

Interpretation of a Raman Spectrum

All vibrational spectroscopies characterize molecular vibrations and to a smaller extent also molecular rotations. Molecular vibrations are based on the movements of the individual atoms of the molecule relative to each other. The forces keeping the molecule together will act like small springs connecting the atoms as illustrated in figure. The set of vibrations is highly dependent on the exact structure of the molecule and therefore comprise a unique vibrational spectrum. This makes vibrational spectroscopy an ideal tool for substance identification. In fact, the vibrational spectrum is so unique that it (or more precisely part of it) is often referred to as the "chemical fingerprint" of the molecule. Different vibrational spectroscopies can detect a different subset of the full vibrational spectrum, which is why the most common methods in this class, Raman and (FT-)IR, are often referred to as "complementary methods".

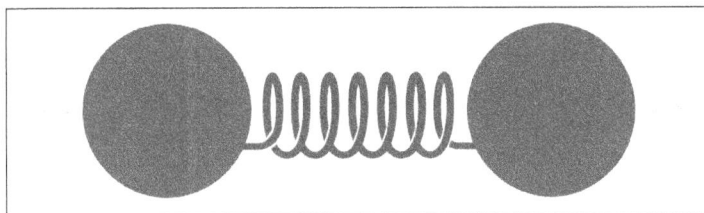

Figure: Vibration of a diatomic molecule in a simplified representation.

Prerequisites for a Molecule to be Raman-active

Raman spectroscopy detects changes in the polarizability of a vibration. It therefore only detects vibrations where the polarizability changes during the movement (these are Raman-active vibrations).

- Polarizability describes how easily the electron cloud around a molecule can be distorted.

- A change in polarizability is, for example, caused by an increase in the size of the electron cloud.

- One example of a vibrational motion increasing the size of the (local) electron cloud of a molecule is a symmetric stretching vibration.

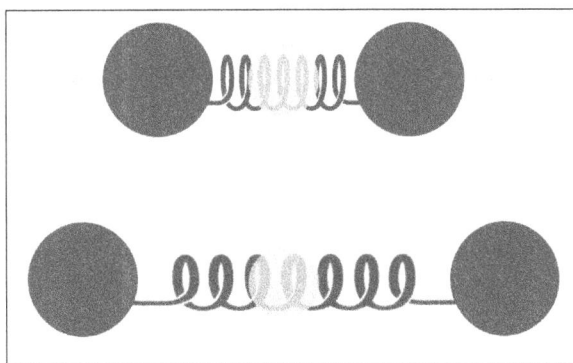

Figure: The symmetric stretching vibration of carbon dioxide (CO_2) increases the size of the electron cloud. It is therefore Raman-active.

Visualization of Raman Spectra

Commonly accepted practice is to plot Raman spectra in a plot "Count Rate" vs. "Raman Shift".

The Raman shift is the energy difference between the incident (laser) light and the scattered (detected) light. This difference is then only connected to the energetic properties of the molecular vibrations studied and hence independent of the laser wavelength. The Raman shift is usually expressed in wavenumbers.

The count rate is the number of events the detector registers for the respective Raman shift per second of detector integration. It is proportional to the intensity of the light imaged to the detector.

Interpretation of Raman Spectra

There are a number of approaches which can be used to interpret Raman spectra, including the three described below:

By Identifying Functional Groups within Molecules

The vibrations of certain distinct subunits of a molecule, called its functional groups, will appear in a Raman spectrum at characteristic Raman shifts. Such a shift is similar for all molecules containing the same functional group. These signals are particularly useful when monitoring reactions which involve these functional groups (oxidation, polymerization, etc.), since they provide a direct measure of the progress during the reaction.

Using these characteristic shifts makes it possible to relate the spectrum of an unknown compound to a class of substances, for example the stretching vibration of the carbonyl group in an aldehyde is always in the range of 1730 cm^{-1} to 1700 cm^{-1}. Figure demonstrates the Raman spectrum of benzonitrile containing the stretching vibration of the cyano-group (CN) of benzonitrile at a characteristic value of 2229.4 cm^{-1}.

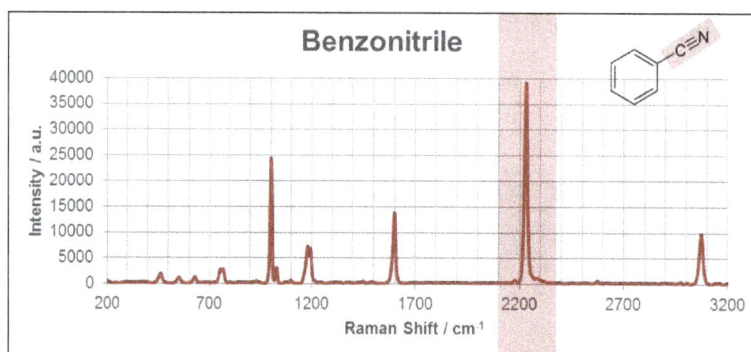

Figure: Raman spectrum of benzonitrile and the stretching vibration of the cyano-group (CN) of benzonitrile at 2229.4 cm^{-1} (red).

Using the Characteristic "Fingerprint Region"

Apart from the molecular vibrations of specific functional groups, vibrations of the molecular scaffolding (skeletal vibrations) can be detected in a Raman spectrum. Skeletal vibrations are usually found at Raman shifts below 1500 cm^{-1} and have a substance-specific, characteristic pattern. This region, often referred to as the "fingerprint" region of a substance, is the most important part of the spectrum for identification purposes.

With Computer-aided Interpretation Software

Substance identification using Raman spectroscopy is nowadays carried out by using software containing a comparison algorithm and a spectral database. The result is shown as a matching factor – the Hit Quality Index (HQI). This factor ranges from 0 (for "no match") to 100 (for "exact match"). The user may then define a threshold down to which the HQI is interpreted as a match. In this way substance identification is possible within seconds and non-technical users can easily interpret the results.

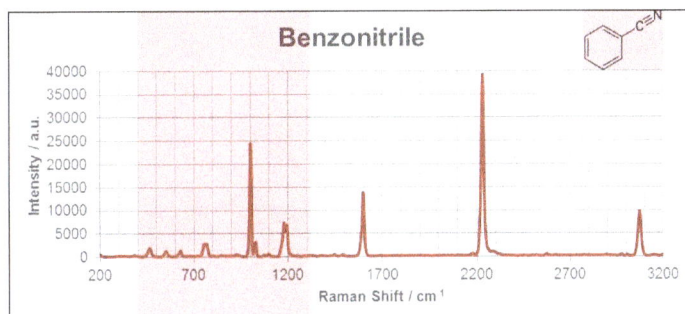

Figure: Raman spectrum of benzonitrile and the skeletal vibration region (chemical fingerprint region) (red).

Raman Scattering

When light encounters molecules in the air, the predominant mode of scattering is elastic scattering, called Rayleigh scattering. This scattering is responsible for the blue color of the sky; it increases with the fourth power of the frequency and is more effective at short wavelengths. It is also possible for the incident photons to interact with the molecules in such a way that energy is either gained or lost so that the scattered photons are shifted in frequency. Such inelastic scattering is called Raman scattering.

Like Rayleigh scattering, the Raman scattering depends upon the polarizability of the molecules. For polarizable molecules, the incident photon energy can excite vibrational modes of the molecules, yielding scattered photons which are diminished in energy by the amount of the vibrational transition energies. A spectral analysis of the scattered light under these circumstances will reveal spectral satellite lines below the Rayleigh scattering peak at the incident frequency. Such lines are called "Stokes lines". If there is significant excitation of vibrational excited states of the scattering molecules, then it is also possible to observe scattering at frequencies above the incident frequency as the vibrational energy is added to the incident photon energy. These lines, generally weaker, are called anti-Stokes lines.

Although finding some application in vibrational spectroscopy of molecules, the use of direct infrared sources for such spectroscopy is usually much easier. Raman spectroscopy has found some application in remote monitoring for pollutants. For example, the scattering produced by a laser beam directed on the plume from an industrial smokestack can be used to monitor the effluent for levels of molecules which will produce recognizable Raman lines.

Raman scattering can also involve rotational transitions of the molecules from which the scattering occurs. Thornton and Rex picture a photon of energy slightly than the energy separation of two levels being scattered, with the excess energy released in the form of a photon of lower energy. Since this is a two-photon process, the selection rule is $\Delta J = +/-2$ for rotational Raman transitions. The sketch below is an idealized depiction of a Raman line produced by interaction of a photon with a diatomic molecule for which the rotational energy levels depend upon one moment of inertia. The upper electronic state of such a molecule can have different levels of rotational and vibrational energy. In this case the upper state is shown as being in rotational state J with scattering associated with an incoming photon at energy matching the J+2 state.

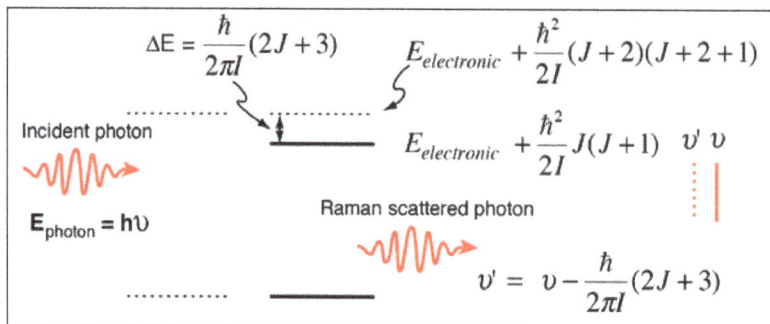

$$\Delta E = \frac{\hbar}{2\pi I}(2J+3) \qquad E_{electronic} + \frac{\hbar^2}{2I}(J+2)(J+2+1)$$

Incident photon

$$E_{electronic} + \frac{\hbar^2}{2I}J(J+1) \quad \upsilon' \; \upsilon$$

$$E_{photon} = h\upsilon$$

Raman scattered photon

$$\upsilon' = \upsilon - \frac{\hbar}{2\pi I}(2J+3)$$

Since the Raman effect depends upon the polarizability of the molecule, it can be observed for molecules which have no net dipole moment and therefore produce no pure rotational spectrum. This process can yield information about the moment of inertia and hence the structure of the molecule.

In Raman scattering, an intense monochromatic light source (laser) can give scattered light which includes one or more "sidebands" that are offset by rotational and/or vibrational energy differences. This is potentially very useful for remote sensing. The mechanism of Raman scattering is different from that of infrared absorption, and Raman and IR spectra provide complementary information.

Typical applications are in structure determination, multicomponent qualitative analysis, and quantitative analysis.

Theory

The Raman scattering transition moment is:

$$R = \left\langle \psi_i \left| a \right| \psi_j \right\rangle$$

where ψ_i and ψ_j are the initial and final states, respectively, and a is the polarizability of the molecule:

$$a = a_0 + (r - r_e)\left(\frac{da}{dr}\right) + \dots \text{ higher terms}$$

Where r is the distance between atoms and ao is the polarizability at the equilibrium bond length, re.

Polarizability can be defined as the ease of which an electron cloud can be distorted by an external electric field.

Since ao is a constant and $\left\langle \psi_i | \psi_j \right\rangle = 0$ R simplifies to:

$$R = \left\langle \psi_i \left| (r - r_e)\left(\frac{da}{dr}\right) \right| \psi_j \right\rangle$$

The result is that there must be a change in polarizability during the vibration for that vibration to inelastically scatter radiation.

Examples of Raman active and inactive vibrations in CO_2.

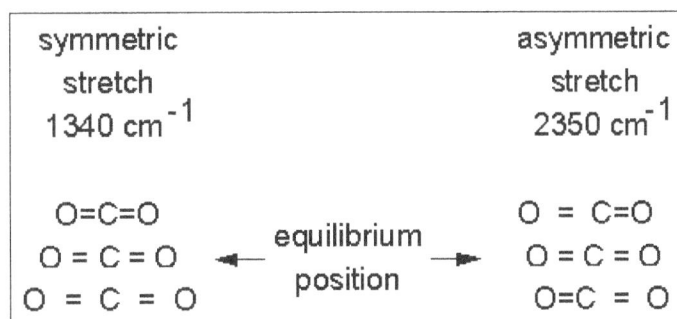

```
  symmetric                          asymmetric
    stretch                            stretch
  1340 cm⁻¹                          2350 cm⁻¹


   O=C=O                              O  =  C=O
   O = C = O      ⟵  equilibrium  ⟶   O = C = O
   O  =  C  =  O        position      O=C  =  O
```

The polarizability depends on how tightly the electrons are bound to the nuclei. In the symmetric stretch the strength of electron binding is different between the minimum and maximum inter-nuclear distances. Therefore the polarizability changes during the vibration and this vibrational mode scatters Raman light (the vibration is Raman active).

In the asymmetric stretch the electrons are more easily polarized in the bond that expands but are less easily polarized in the bond that compresses.

There is no overall change in polarizability and the asymmetric stretch is Raman inactive.

Raman line intensities are proportional to: $v4 * \sigma(v) * I * (-Ei/kT) * C$ where where v is the frequency of the incident radiation, $\sigma(v)$) is the Raman cross section (typically 10-29 cm²), I is the radiation intensity, exp(-Ei/kT) is the Boltzmann factor for state i, and C is the analyte concentration.

Instrumentation

The most common light source in Raman spectroscopy is an Ar-ion laser. Resonance Raman spectroscopy requires tunable radiation and sources are Ar-ion-laser-pumped cw dye lasers, or high-repetition-rate excimer-laser-pumped pulsed dye lasers. Because Raman scattering is a weak process, a key requirement to obtain Raman spectra is that thespectrometer provide a high rejection of scattered laser light.

New methods such as very narrow rejection filters and Fourier-transform techniques are becoming more widespread.

Infrared Spectroscopy

Infrared spectroscopy (IR) is a characterization tool chemists use to help determine the molecular structure. IR capitalizes on the concept that functional groups absorb specific frequencies of energy based on their structure. When a functional group absorbs energy, it can vibrate in a bending or stretching mode and the characteristic energy for this vibrational mode is reported in wavenumbers. A change in molecular dipole is required for a bending or stretching mode to be visible in by IR. In general, stronger bonds vibrate at higher wavenumbers, as do bonds between atoms of very different size. IR is a reliable and sensitive technique that has important applications in crime-scene investigation, for example in solving a case of arson, IR can identify traces of fuel or accelerant.

The IR spectroscopy concept can generally be analyzed in three ways: by measuring reflection, emission, and absorption. The major use of infrared spectroscopy is to determine the functional groups of molecules, relevant to both organic and inorganic chemistry.

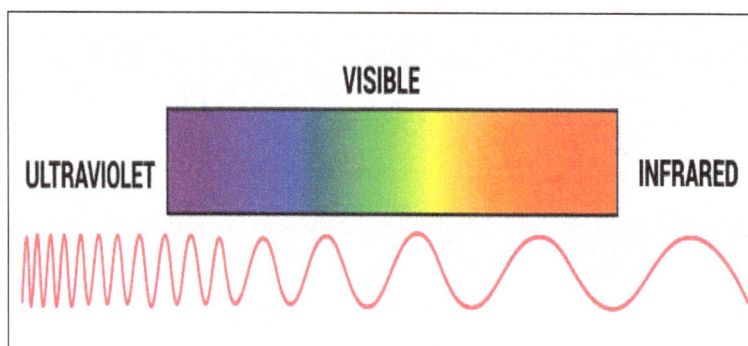

An IR spectrum is essentially a graph plotted with the infrared light absorbed on the Y-axis against. frequency or wavelength on the X-axis. An illustration highlighting the different regions that light can be classified into is given below.

IR Spectroscopy detects frequencies of infrared light that are absorbed by a molecule. Molecules tend to absorb these specific frequencies of light since they correspond to the frequency of the vibration of bonds in the molecule.

The energy required to excite the bonds belonging to a molecule, and to make them vibrate with more amplitude, occurs in the Infrared region. A bond will only interact with the electromagnetic infrared radiation, however, if it is polar.

The presence of separate areas of partial positive and negative charge in a molecule allows the electric field component of the electromagnetic wave to excite the vibrational energy of the molecule.

The change in the vibrational energy leads to another corresponding change in the dipole moment of the given molecule. The intensity of the absorption depends on the polarity of the bond. Symmetrical non-polar bonds in N=N and O=O do not absorb radiation, as they cannot interact with an electric field.

Regions of the Infrared Spectrum

Most of the bands that indicate what functional group is present are found in the region from 4000 cm^{-1} to 1300 cm^{-1}. Their bands can be identified and used to determine the functional group of an unknown compound.

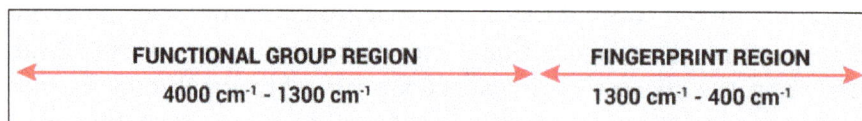

Bands that are unique to each molecule, similar to a fingerprint, are found in the fingerprint region, from 1300 cm^{-1} to 400 cm^{-1}. These bands are only used to compare the spectra of one compound to another.

Samples in Infrared Spectroscopy

The samples used in IR spectroscopy can be either in the solid, liquid, or gaseous state.

- Solid samples can be prepared by crushing the sample with a mulling agent which has an oily texture. A thin layer of this mull can now be applied on a salt plate to be measured.

- Liquid samples are generally kept between two salt plates and measured since the plates are transparent to IR light. Salt plates can be made up of sodium chloride, calcium fluoride, or even potassium bromide.

- Since the concentration of gaseous samples can be in parts per million, the sample cell must have a relatively long pathlength, i.e. light must travel for a relatively long distance in the sample cell.

Thus, samples of multiple physical states can be used in Infrared Spectroscopy.

Principle of Infrared Spectroscopy

The IR spectroscopy theory utilizes the concept that molecules tend to absorb specific frequencies of light that are characteristic of the corresponding structure of the molecules. The energies are reliant on the shape of the molecular surfaces, the associated vibronic coupling, and the mass corresponding to the atoms.

For instance, the molecule can absorb the energy contained in the incident light and the result is a faster rotation or a more pronounced vibration.

IR Spectroscopy Instrumentation

The instrumentation of infrared spectroscopy is illustrated below. First, a beam of IR light from the source is split into two and passed through the reference ant the sample respectively.

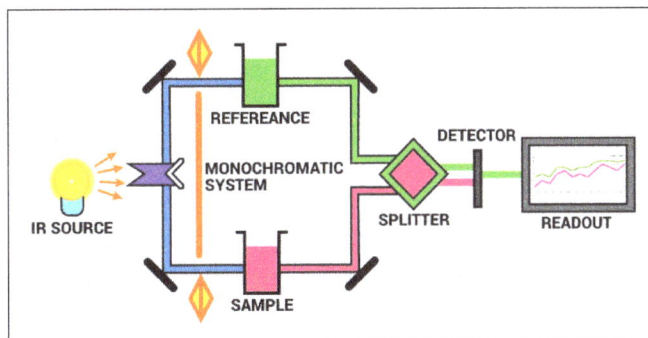

Now, both of these beams are reflected to pass through a splitter and then through a detector. Finally, the required reading is printed out after the processor deciphers the data passed through the detector.

X-ray Spectroscopy

X-ray spectroscopy is a technique that detects and measures photons, or particles of light, that have wavelengths in the X-ray portion of the electromagnetic spectrum. It's used to help scientists understand the chemical and elemental properties of an object.

There are several different X-ray spectroscopy methods that are used in many disciplines of science and technology, including archaeology, astronomy and engineering. These methods can be used independently or together to create a more complete picture of the material or object being analyzed.

Working of X-ray Spectroscopy

When an atom is unstable or is bombarded with high-energy particles, its electrons transition from one energy level to another. As the electrons adjust, the element absorbs and releases high-energy X-ray photons in a way that's characteristic of atoms that make up that particular chemical element. X-ray spectroscopy measures those changes in energy, which allows scientists to identify elements and understand how the atoms within various materials interact.

There are two main X-ray spectroscopy techniques: wavelength-dispersive X-ray spectroscopy (WDXS) and energy-dispersive X-ray spectroscopy (EDXS). WDXS measures the X-rays of a single wavelength that are diffracted by a crystal. EDXS measures the X-ray radiation emitted by electrons stimulated by a high-energy source of charged particles.

In both techniques, how the radiation is dispersed indicates the atomic structure of the material and therefore, the elements within the object being analyzed.

Selection Rules of Spectroscopy

A selection rule describes how the probability of transitioning from one level to another cannot be zero. It has two sub-pieces: a gross selection rule and a specific selection rule. A gross selection rule illustrates characteristic requirements for atoms or molecules to display a spectrum of a given kind, such as an IR spectroscopy or a microwave spectroscopy. Once the atom or molecules follow the gross selection rule, the specific selection rule must be applied to the atom or molecules to determine whether a certain transition in quantum number may happen or not.

Selection rules specify the possible transitions among quantum levels due to absorption or emission of electromagnetic radiation. Incident electromagnetic radiation presents an oscillating electric field $E_0 \cos(\omega t)$ that interacts with a transition dipole. The dipole operator is $\mu = e \cdot r$ where r is a vector pointing in a direction of space.

A dipole moment of a given state is,

$$\mu_z = \int \Psi_1 * \mu_z \Psi_1 \, d\tau$$

A transition dipole moment is a transient dipolar polarization created by an interaction of electromagnetic radiation with a molecule,

$$(\mu_z)_{12} = \int \Psi_1 * \mu_z \Psi_2 \, d\tau$$

In an experiment we present an electric field along the z axis (in the laboratory frame) and we may consider specifically the interaction between the transition dipole along the x, y, or z axis of the molecule with this radiation. If μ_z is zero then a transition is forbidden. The selection rule is a statement of when μ_z is non-zero.

We can consider selection rules for electronic, rotational, and vibrational transitions.

Electronic Transitions

We consider a hydrogen atom. In order to observe emission of radiation from two states muz must be non-zero. That is,

$$(\mu_z)_{12} = \int \Psi_1^* e \cdot z \Psi_2 \, d\tau \neq 0$$

For example, is the transition from Ψ_{1s} to Ψ_{2s} allowed,

$$(\mu_z)_{12} = \int \Psi_{1s}^* e \cdot z \Psi_{2s} \, d\tau$$

Using the fact that z = r cosq in spherical polar coordinates we have,

$$(\mu_z)_{12} = e \iiint e^{-r/a_0} r \cos\theta \left(2 - \frac{r}{a_0}\right) e^{-r/a_0} r^2 \sin\theta \, dr d\theta \, d\phi$$

We can consider each of the three integrals separately,

$$\int_0^\infty e^{-r/a_0} r \left(2 - \frac{r}{a_0}\right) e^{-r/a_0} r^2 dr \int_0^\pi \cos\theta \sin\theta d\theta \int_0^{2\pi} d\phi$$

If any one of these is non-zero the transition is not allowed. We can see specifically that we should consider the q integral. We make the substitution x=cos q, d x=−sin q, d and the integral becomes which is zero.

$$-\int_1^{-1} x dx = -\frac{x^2}{2}\bigg|_1^{-1} = 0$$

The result is an even function evaluated over odd limits. In a similar fashion we can show that transitions along the x or y axes are not allowed either. This presents a selection rule that transitions are forbidden for $\Delta l = 0$. For electronic transitions the selection rules turn out to be $\Delta l = \pm 1$ and $\Delta m = 0$ These result from the integrals over spherical harmonics which are the same for rigid rotator wavefunctions. We will prove the selection rules for rotational transitions keeping in mind that they are also valid for electronic transitions.

Rotational Transitions

We can use the definition of the transition moment and the spherical harmonics to derive selection rules for a rigid rotator. Once again we assume that radiation is along the z axis.

Notice that m must be non-zero in order for the transition moment to be non-zero. This proves that a molecule must have a permanent dipole moment in order to have a rotational spectrum. The spherical harmonics can be written as,

$$Y_J^M(\theta,\phi) = N_{JM}P_J^{|M|}(cos\theta)e^{iM\phi}$$

where N_{JM} is a normalization constant. Using the standard substitution of $x = \cos q$ we can express the rotational transition moment as,

$$(\mu_z)_{J,M,J',M'} = \mu_N J_M N_{J'M'} \int_0^{2\pi} e^{I(M-M')\phi}d\phi \int_{-1}^1 P_{J'}^{|M'|}(x)P_J^{|M|}(x)dx$$

The integral over f is zero unless M = M' so ΔM= 0 is part of the rigid rotator selection rule. Integration over φ for M=M' gives 2π so we have,

$$(\mu_z)_{J,M,J',M'} = 2\pi\mu N_{JM}N_{J'M'} \int_{-1}^1 P_{J'}^{|M'|}(x)P_J^{|M|}(x)dx$$

We can evaluate this integral using the identity,

$$(2J+1)(x)P_J^{|M|}(x) = (J-|M|+1)P_{J+1}^{|M|}(x) + (J-|M|)P_{J+1}^{|M|}(x)$$

Substituting into the integral one obtains an integral which will vanish unless , $J' = J + 1$ or $J' = J - 1$,

$$\int_{-1}^1 P_{J'}^{|M'|}(x)\left(\frac{(J-|M|+1)}{(2J+1)}P_{J+1}^{|M|}(x) + \frac{(J-|M|)}{(2J+1)}P_{J-1}^{|M|}(x)\right)dx$$

This leads to the selection rule $\Delta J = \pm 1$ for absorptive rotational transitions. Keep in mind the physical interpretation of the quantum numbers J and M as the total angular momentum and z-component of angular momentum, respectively. These selection rules also apply to the orbital angular momentum ($\Delta l = \pm 1, \Delta m = 0$).

Vibrational Transitions

The harmonic oscillator wavefunctions are,

$$\Psi_v(q) = N_v H_v(\alpha^{1/2}q)e^{-\alpha q^2/2}$$

where $H_v(a1/2q)$ is a Hermite polynomial and a = (km/á2)1/2.

he transition dipole moment for electromagnetic radiation polarized along the z axis is,

$$(\mu_z)_{v,v'} = \int_{-\infty}^{\infty} N_v N_{v'} H_{v'}(\alpha^{1/2}q)e^{-\alpha q^2/2} H \mu_z(\alpha^{1/2}q)e^{-\alpha q^2/2}dq$$

Note that we continue to use the general coordinate q although this can be z if the dipole moment of the molecule is aligned along the z axis. The transition moment can be expanded about the equilibrium nuclear separation.

$$\mu_z(q) = \mu_0 + \left(\frac{\partial \mu}{\partial q}\right) q + ...$$

where mo is the dipole moment at the equilibrium bond length and q is the displacement from that equilibrium state. From the first two terms in the expansion we have for the first term,

$$(\mu_z)_{v,v'} = \mu_0 \int_{-\infty}^{\infty} N_v N_{v'} H_{v'}(\alpha^{1/2}q)e^{-\alpha q^2/2} H_v(\alpha^{1/2}q)e^{-\alpha q^2/2} dq$$

This term is zero unless v = v' and in that case there is no transition since the quantum number has not changed.

$$(\mu_z)_{v,v'} = \left(\frac{\partial \mu}{\partial q}\right) \int_{-\infty}^{\infty} N_v N_{v'} H_{v'}(\alpha^{1/2}q)e^{-\alpha q^2/2} H_v(\alpha^{1/2}q)e^{-\alpha q^2/2} dq$$

This integral can be evaluated using the Hermite polynomial identity known as a recursion relation,

$$xH_v(x) = vH_{v-1}(x) + \frac{1}{2}H_{v+1}(x)$$

where x = Öaq. If we now substitute the recursion relation into the integral we find:

$$(\mu_z)_{v,v'} = \frac{N_v N_{v'}}{\sqrt{a}}\left(\frac{\partial \mu}{\partial q}\right)$$

$$\int_{-\infty}^{\infty} H_{v'}(\alpha^{1/2}q)e^{-\alpha q^2/2}\left(vH_{v-1}(\alpha^{1/2}q) + \frac{1}{2}H_{v+1}(\alpha^{1/2}q)\right)dq$$

which will be non-zero if v' = v − 1 or v' = v + 1. Thus, we see the origin of the vibrational transition selection rule that v = ± 1. We also see that vibrational transitions will only occur if the dipole moment changes as a function nuclear motion.

Electromagnetic Radiation

Electromagnetic radiation is a form of energy that is produced by oscillating electric and magnetic disturbance, or by the movement of electrically charged particles traveling through a vacuum or matter. The electric and magnetic fields come at right angles to each other and combined wave moves perpendicular to both magnetic and electric oscillating fields thus the disturbance. Electron radiation is released as photons, which are bundles of light energy that travel at the speed of light as quantized harmonic waves. This energy is then grouped into categories based on its wavelength into the electromagnetic spectrum. These electric and magnetic waves travel perpendicular to each other and have certain characteristics, including amplitude, wavelength, and frequency.

General Properties of all electromagnetic radiation:

1. Electromagnetic radiation can travel through empty space. Most other types of waves must travel through some sort of substance. For example, sound waves need either a gas, solid, or liquid to pass through in order to be heard.

2. The speed of light is always a constant.

3. Wavelengths are measured between the distances of either crests or troughs. It is usually characterized by the Greek symbol λ.

Waves and their Characteristics

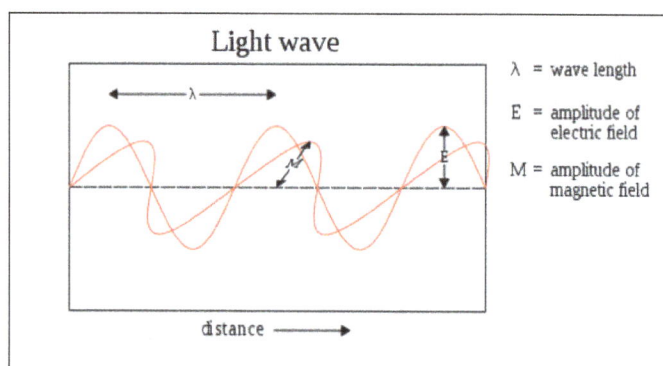

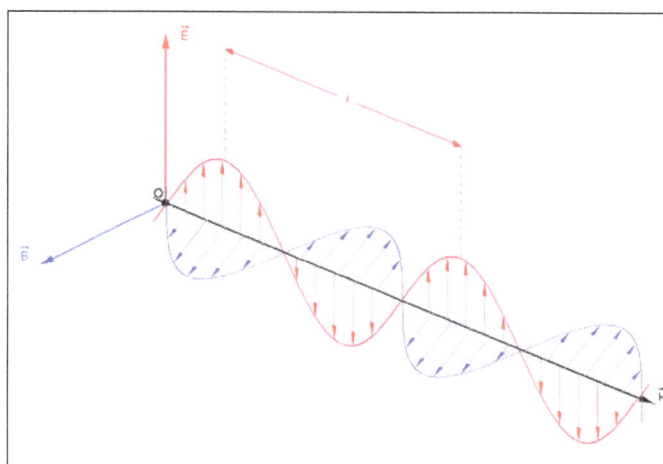

Figure: Electromagnetic Waves.

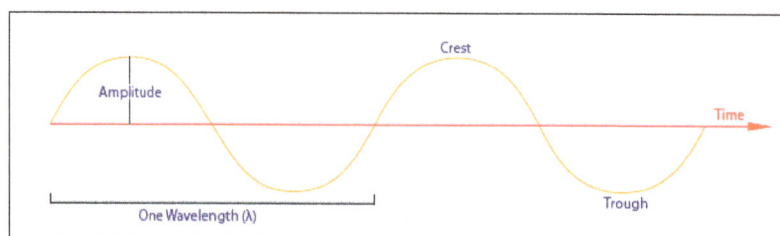

Figure: An EM Wave.

Amplitude

Amplitude is the distance from the maximum vertical displacement of the wave to the middle of the wave. This measures the magnitude of oscillation of a particular wave. In short, the amplitude is basically the height of the wave. Larger amplitude means higher energy and lower amplitude means lower energy. Amplitude is important because it tells you the intensity or brightness of a wave in comparison with other waves.

Wavelength

Wavelength (λ) is the distance of one full cycle of the oscillation. Longer wavelength waves such as radio waves carry low energy; this is why we can listen to the radio without any harmful consequences. Shorter wavelength waves such as x-rays carry higher energy that can be hazardous to our health. Consequently lead aprons are worn to protect our bodies from harmful radiation when we undergo x-rays. This wavelength frequently relationship is characterized by:

$$c = \lambda v.$$

Where,

- c is the speed of light,

- λ is wavelength, and

- v is frequency.

Shorter wavelength means greater frequency, and greater frequency means higher energy. Wavelengths are important in that they tell one what type of wave one is dealing with.

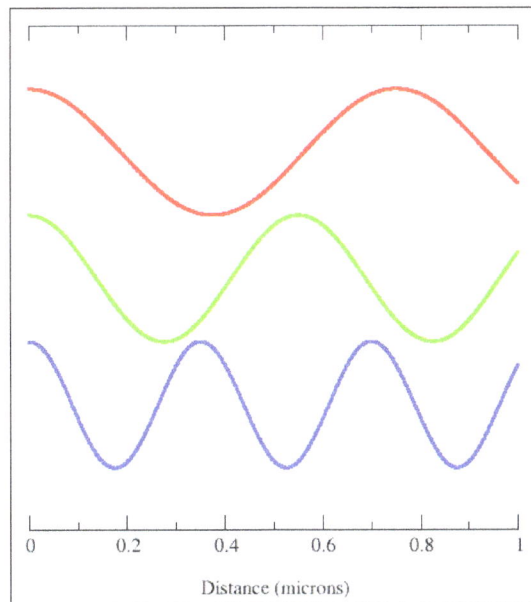

Figure: Different Wavelengths and Frequencies.

Remember, Wavelength tells you the type of light and Amplitude tells you about the intensity of the light.

Frequency

Frequency is defined as the number of cycles per second, and is expressed as sec^{-1} or Hertz (Hz). Frequency is directly proportional to energy and can be express as:

$$E = h\nu$$

Where,

- E is energy,

- h is Planck's constant, (h= 6.62607 x 10^{-34} J), and

- ν is frequency.

Period

Period (T) is the amount of time a wave takes to travel one wavelength; it is measured in seconds (s).

Velocity

The velocity of wave in general is expressed as:

$$velocity = \lambda \nu$$

For Electromagnetic wave, the velocity in vacuum is $2.99 \times 10^8 \ m/s$ miles/second.

Radiation Types

Radio Waves: It is approximately 10^3 m in wavelength. As the name implies, radio waves are transmitted by radio broadcasts, TV broadcasts, and even cell phones. Radio waves have the lowest energy levels. Radio waves are used in remote sensing, where hydrogen gas in space releases radio energy with a low frequency and is collected as radio waves. They are also used in radar systems, where they release radio energy and collect the bounced energy back. Especially useful in weather, radar systems are used to can illustrate maps of the surface of the Earth and predict weather patterns since radio energy easily breaks through the atmosphere.

Microwaves: It can be used to broadcast information through space, as well as warm food. They are also used in remote sensing in which microwaves are released and bounced back to collect information on their reflections.

Microwaves can be measured in centimeters. They are good for transmitting information because the energy can go through substances such as clouds and light rain. Short microwaves are sometimes used in Doppler radars to predict weather forecasts.

Infrared radiation: It can be released as heat or thermal energy. It can also be bounced back, which is called near infrared because of its similarities with visible light energy. Infrared Radiation is most commonly used in remote sensing as infrared sensors collect thermal energy, providing us with weather conditions.

This picture represents a snap shot in mid-infrared light.

Visible Light: It is the only part of the electromagnetic spectrum that humans can see with an unaided eye. This part of the spectrum includes a range of different colors that all represent a particular wavelength. Rainbows are formed in this way; light passes through matter in which it is absorbed or reflected based on its wavelength. Thus, some colors are reflected more than other, leading to the creation of a rainbow.

Table: The color regions of the Visible Spectrum.

Color Region	Wavelength (nm)
Violet	380-435
Blue	435-500
Cyan	500-520
Green	520-565
Yellow	565-590
Orange	590-625
Red	625-740

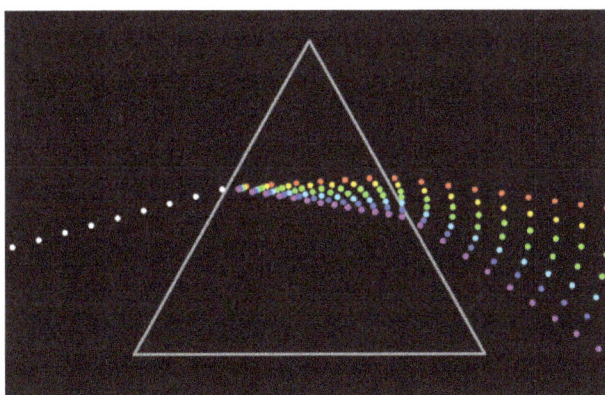

Figure: Dispersion of Light Through A Prism.

Ultraviolet, Radiation, X-Rays, and Gamma Rays are all related to events occurring in space. UV radiation is most commonly known because of its severe effects on the skin from the sun, leading to cancer. X-rays are used to produce medical images of the body. Gamma Rays can used in chemotherapy in order to rid of tumors in a body since it has such a high energy level. The shortest waves, Gamma rays, are approximately 10^{-12} m in wavelength. Out this huge spectrum, the human eyes can only detect waves from 390 nm to 780 nm.

Equations of Waves

The mathematical description of a wave is:

$$y = A \sin(kx - \omega t)$$

where A is the amplitude, k is the wave number, x is the displacement on the x-axis.

$$k = \frac{2\pi}{\lambda}$$

where λ is the wavelength. Angular frequency described as:

$$\omega = 2\pi v = \frac{2\pi}{T}$$

where v is frequency and period (T) is the amount of time for the wave to travel one wavelength.

Interference

An important property of waves is the ability to combine with other waves. There are two type of interference: constructive and destructive. Constructive interference occurs when two or more waves are in phase and and their displacements add to produce a higher amplitude. On the contrary, destructive interference occurs when two or more waves are out of phase and their displacements negate each other to produce lower amplitude.

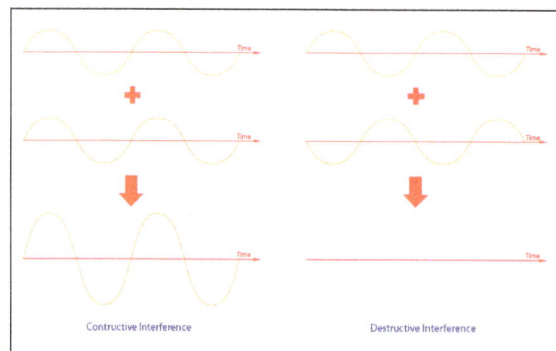

Figure: Constructive and Destructive Interference.

Interference can be demonstrated effectively through the double slit experiment. This experiment consists of a light source pointing toward a plate with one slit and a second plate with two slits. As the light travels through the slits, we notice bands of alternating intensity on the wall behind the second plate. The banding in the middle is the most intense because the two waves are perfectly in phase at that point and thus constructively interfere. The dark bands are caused by out of phase waves which result in destructive interference. This is why you observe nodes on figure. In a similar way, if electrons are used instead of light, electrons will be represented both as waves and particles.

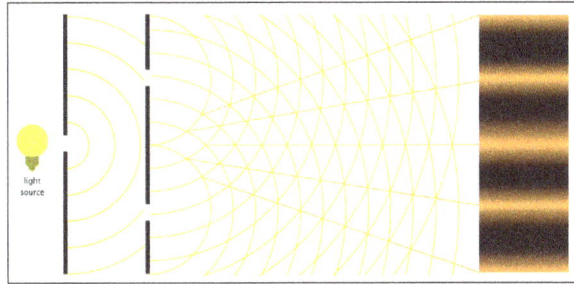

Figure: Double-Slit Interference Experiment.

Wave-particle Duality

Electromagnetic radiation can either acts as a wave or a particle, a photon. As a wave, it is represented by velocity, wavelength, and frequency. Light is an EM wave since the speed of EM waves is the same as the speed of light. As a particle, EM is represented as a photon, which transports energy. When a photon is absorbed, the electron can be moved up or down an energy level. When it moves up, it absorbs energy, when it moves down, energy is released. Thus, since each atom has its own distinct set of energy levels, each element emits and absorbs different frequencies. Photons with higher energies produce shorter wavelengths and photons with lower energies produce longer wavelengths.

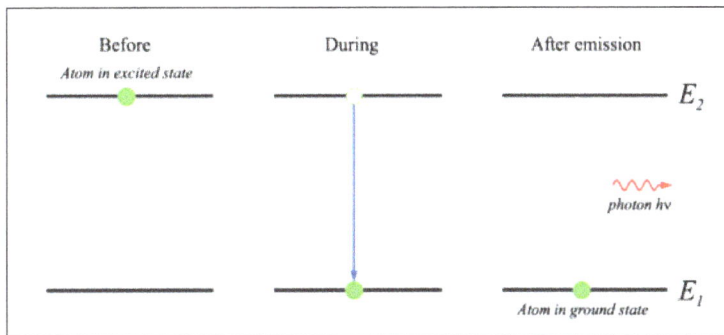

Figure: Photon Before and After Emission.

Ionizing and Non-ionizing Radiation

Electromagnetic Radiation is also categorized into two groups based, ionizing and non-ionizing, on the severity of the radiation. Ionizing radiation holds a great amount of energy to remove electrons and cause the matter to become ionized. Thus, higher frequency waves such as the X-rays and gamma-rays have ionizing radiation. However, lower frequency waves such as radio waves, do not have ionizing radiation and are grouped as non-ionizing.

Electromagnetic Radiation and Temperature

Electromagnetic radiation released is related to the temperature of the body. Stephan-Boltzmann Law says that if this body is a black body, one which perfectly absorbs and emits radiation, the radiation released is equal to the temperature raised to the fourth power. Therefore, as temperature increases, the amount of radiation released increases greatly. Objects that release radiation very well also absorb radiation at certain wavelengths very well. This is explained by the Kirchhoff's Law. Wavelengths are also related to temperature. As the temperature increases, the wavelength of maximum emission decreases.

Electromagnetic Spectrum

Electromagnetic spectrum is the entire distribution of electromagnetic radiation according to frequency or wavelength. Although all electromagnetic waves travel at the speed of light in a vacuum, they do so at a wide range of frequencies, wavelengths, and photon energies. The electromagnetic spectrum comprises the span of all electromagnetic radiation and consists of many subranges, commonly referred to as portions, such as visible light or ultraviolet radiation. The various portions bear different names based on differences in behaviour in the emission, transmission, and absorption of the corresponding waves and also based on their different practical applications. There are no precise accepted boundaries between any of these contiguous portions, so the ranges tend to overlap.

The entire electromagnetic spectrum, from the lowest to the highest frequency (longest to shortest wavelength), includes all radio waves (e.g., commercial radio and television, microwaves, radar), infrared radiation, visible light, ultraviolet radiation, X-rays, and gamma rays. Nearly all frequencies and wavelengths of electromagnetic radiation can be used for spectroscopy.

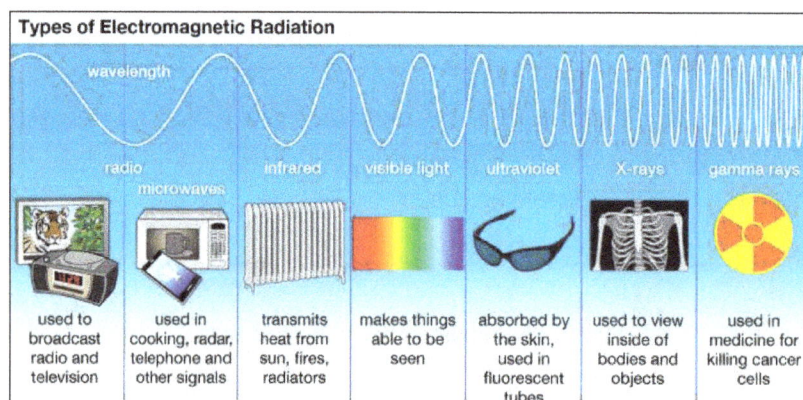

Figure: Radio waves, infrared rays, visible light, ultraviolet rays, X-rays, and gamma rays
are all types of electromagnetic radiation. Radio waves have the longest
wavelength, and gamma rays have the shortest wavelength.

The electromagnetic (EM) spectrum is the range of all types of EM radiation. Radiation is energy that travels and spreads out as it goes – the visible light that comes from a lamp in your house and the radio waves that come from a radio station are two types of electromagnetic radiation. The other types of EM radiation that make up the electromagnetic spectrum are microwaves, infrared light, ultraviolet light, X-rays and gamma-rays.

The image below shows where you might encounter each portion of the EM spectrum in your day-to-day life.

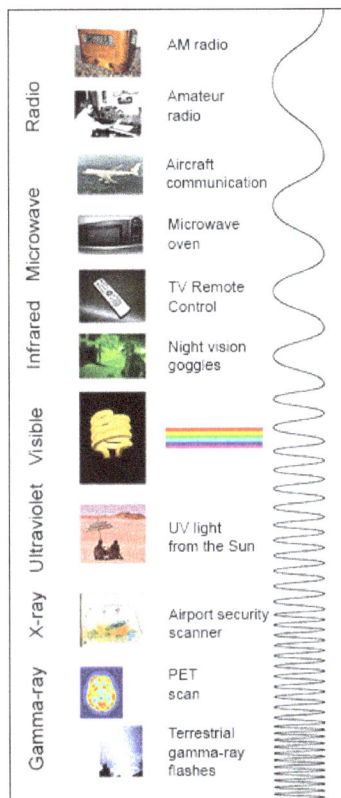

Figure: The electromagnetic spectrum from lowest energy/longest wavelength
(at the top) to highest energy/shortest wavelength (at the bottom).

- Radio: Your radio captures radio waves emitted by radio stations, bringing your favorite tunes. Radio waves are also emitted by stars and gases in space.

- Microwave: Microwave radiation will cook your popcorn in just a few minutes, but is also used by astronomers to learn about the structure of nearby galaxies.

- Infrared: Night vision goggles pick up the infrared light emitted by our skin and objects with heat. In space, infrared light helps us map the dust between stars.

- Visible: Our eyes detect visible light. Fireflies, light bulbs, and stars all emit visible light.

- Ultraviolet: Ultraviolet radiation is emitted by the Sun and are the reason skin tans and burns. Hot objects in space emit UV radiation as well.

- X-ray: A dentist uses X-rays to image your teeth, and airport security uses them to see through your bag. Hot gases in the Universe also emit X-rays.

- Gamma ray: Doctors use gamma-ray imaging to see inside your body. The biggest gamma-ray generator of all is the Universe.

Radio Wave and Gamma Ray

Are radio waves completely different physical objects than gamma-rays? They are produced in different processes and are detected in different ways, but they are not fundamentally different.

Radio waves, gamma-rays, visible light, and all the other parts of the electromagnetic spectrum are electromagnetic radiation.

Electromagnetic radiation can be described in terms of a stream of mass-less particles, called photons, each traveling in a wave-like pattern at the speed of light. Each photon contains a certain amount of energy. The different types of radiation are defined by the the amount of energy found in the photons. Radio waves have photons with low energies, microwave photons have a little more energy than radio waves, infrared photons have still more, then visible, ultraviolet, X-rays, and, the most energetic of all, gamma-rays.

Measuring Electromagnetic Radiation

Electromagnetic radiation can be expressed in terms of energy, wavelength, or frequency. Frequency is measured in cycles per second, or Hertz. Wavelength is measured in meters. Energy is measured in electron volts. Each of these three quantities for describing EM radiation are related to each other in a precise mathematical way. But why have three ways of describing things, each with a different set of physical units? The short answer is that scientists don't like to use numbers any bigger or smaller than they have to. It is much easier to say or write "two kilometers" than "two thousand meters." Generally, scientists use whatever units are easiest for the type of EM radiation they work with.

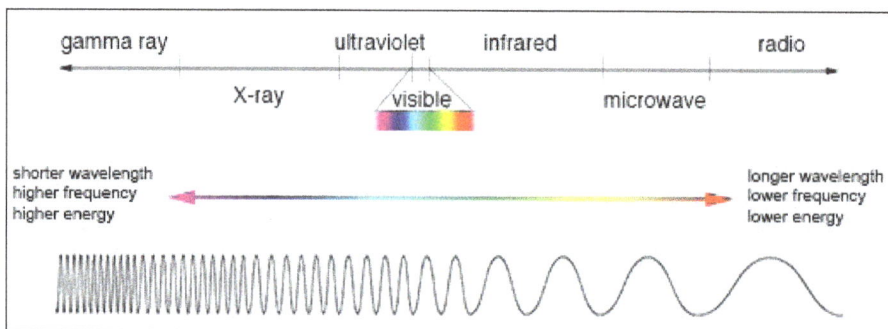

Comparison of wavelength, frequency and energy for the electromagnetic spectrum.

Astronomers who study radio waves tend to use wavelengths or frequencies. Most of the radio part of the EM spectrum falls in the range from about 1 cm to 1 km, which is 30 gigahertz (GHz) to 300 kilohertz (kHz) in frequencies. The radio is a very broad part of the EM spectrum.

Infrared and optical astronomers generally use wavelength. Infrared astronomers use microns (millionths of a meter) for wavelengths, so their part of the EM spectrum falls in the range of 1 to 100 microns. Optical astronomers use both angstroms (0.00000001 cm, or 10^{-8} cm) and nanometers (0.0000001 cm, or 10^{-7} cm). Using nanometers, violet, blue, green, yellow, orange, and red light have wavelengths between 400 and 700 nanometers. (This range is just a tiny part of the entire EM spectrum, so the light our eyes can see is just a little fraction of all the EM radiation around us).

The wavelengths of ultraviolet, X-ray, and gamma-ray regions of the EM spectrum are very small. Instead of using wavelengths, astronomers that study these portions of the EM spectrum usually refer to these photons by their energies, measured in electron volts (eV). Ultraviolet radiation falls in the range from a few electron volts to about 100 eV. X-ray photons have energies in the range 100 eV to 100,000 eV (or 100 keV). Gamma-rays then are all the photons with energies greater than 100 keV.

Reason for Putting Telescopes in Orbit

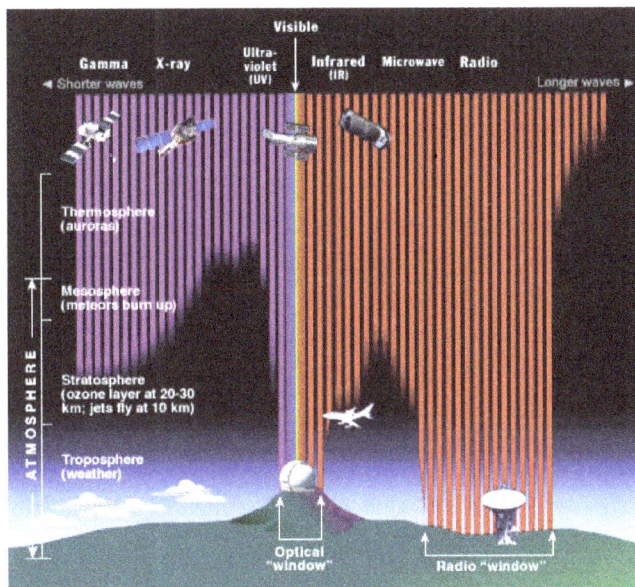

In the above figure, the Earth's atmosphere stops most types of electromagnetic radiation from space from reaching Earth's surface. This illustration shows how far into the atmosphere different parts of the EM spectrum can go before being absorbed. Only portions of radio and visible light reach the surface.

Most electromagnetic radiation from space is unable to reach the surface of the Earth. Radio frequencies, visible light and some ultraviolet light makes it to sea level. Astronomers can observe some infrared wavelengths by putting telescopes on mountain tops. Balloon experiments can reach 35 km above the surface and can operate for months. Rocket flights can take instruments all the way above the Earth's atmosphere, but only for a few minutes before they fall back to Earth.

References

- Spectroscopy, science: britannica.com, Retrieved 15 April, 2019

- What-is-optical-spectroscopy: wisegeek.com, Retrieved 14 June, 2019

- Basics-of-raman-spectroscopy: anton-paar.com, Retrieved 7 July, 2019

- Raman, atmos, hbase: phy-astr.gsu.edu, Retrieved 18 February, 2019

- Raman-spectroscopy: chemicool.com, Retrieved 9 August, 2019

- Infrared-spectroscopy-6: chegg.com, Retrieved 20 February, 2019

- Selection-Rules, Fundamentals-of-Spectroscopy, Spectroscopy, Physical-and-Theoretical-Chemistry: libretexts.org, Retrieved 2 January, 2019

- Electromagnetic-Radiation, Fundamentals-of-Spectroscopy, Spectroscopy, Physical-and-Theoretical-Chemistry: libretexts.org, Retrieved 11 June, 2019

- Electromagnetic-spectrum, science: britannica.com, Retrieved 30 March, 2019

2

Rotational Spectroscopy

Rotational spectroscopy deals with the measurement of the energies of transitions between quantized rotational states of molecules which are in the gas phase. Its most important application is the exploration of the chemical composition of the interstellar medium through radio telescopes. This chapter closely examines the key concepts of rotational spectrometry such as rotation of polyatomic molecules and microwave rotational spectroscopy.

Rotational spectroscopy uses the discrete energy levels of rotation to measure the inertia of the bond and therefore the bond length.

Rotation spectroscopy measures how the rotation of a molecule about its centre mass can give an idea of inertia and therefore bond length. The excitation of a molecule can cause the molecule to start rotating around a point, the energy of this movement is quantised around the point. It is important to note that these transitions can only occur between molecules that do not have an inversion of symmetry. The value for is the reduced mass of the molecule:

$$m_r = \frac{m_1 m_2}{m_1 + m_2}$$

The masses of the molecules have a very real effect on how the molecule rotates and pivots around the molecule. This can give the energy as two masses and then as a subject of inertia:

$$E = \frac{1}{2} m_1 v_1^2 + \frac{1}{2} m_2 v_2^2$$

These two atoms have to move at different speeds, proportional to their distances from the centre of mass otherwise the bond would begin to bend inwards. This is shown by:

$$\frac{v_1}{r_1} = \frac{v_2}{r_2} = \omega$$

And relating this to inertia:

$$E = \frac{1}{2} m_1 r_1^2 \omega^2 + \frac{1}{2} m_1 r_2^2 \omega^2 = \frac{1}{2} I \omega^2$$

The angular momentum can now be related very closely to kinetic energy, this leaves the angular momentum to be given as:

$$L = I \omega$$

The rotational energy can then be written as terms of inertia and angular momentum:

$$E = \frac{L^2}{2I}$$

This only leaves the quantum mechanical result to be achieved by using Schrodinger's equation. Which isn't necessary as we can use the shortcut:

$$L = \frac{h}{2\pi}\sqrt{J(J+1)} \quad J = 0, 1, 2$$

This is where J is the rotational quantum number, the angular momentum that is a vector directed along an axis of rotation with magnitude L. This shows how the value of J is an integer in every molecule, the difference is caused by the change in value of the inertia I.

The energies of this can therefore be given by:

$$E = \frac{h^2}{8\pi^2 I}J(J+1)$$

Or

$$E = BhJ(J+1)$$

As B and h are constants given by:

$$B = \frac{h}{8\pi^2 I}$$

B is the rotational constant of the molecule with units of Hz. The first few rotational energy levels are shown below:

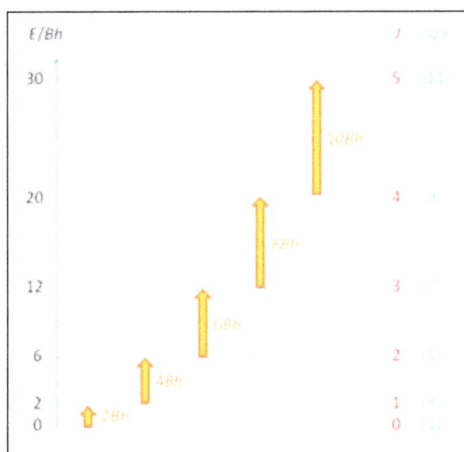

Because the spacing between rotational levels depends on B, which is inversely proportional to the moment of inertia the levels become more and more tightly packed as inertia increases. This is true for both linear and non-linear molecules. The degeneracy of the angular momentum is also important it is quantised and related to quantum mechanical principles also:

$$L_z = M \frac{h}{2\pi} \; where \; M = 0, \pm 1, \pm 2$$

So when J = 2 the total angular momentum is $L_s = \frac{h}{2\pi}\sqrt{6}$ from $L_z = \frac{h}{2p}\sqrt{J(J+1)}$.

The degeneracy causes 2J+1 energy values to form. As the value of L is the angular momentum vector when there is no external force acting on the molecule L is unchanged. So there is no preferred direction when there is no magnetic field present so all values of M have the same energy. When a magnetic field is induced along an axis any extra energy produced needs to be displaced. The vectors can vary by multiples of h/2π. This is shown in the diagram below:

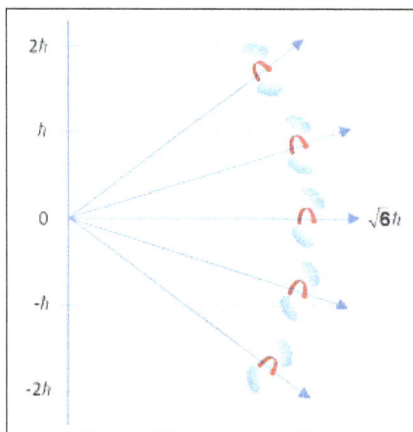

The degeneracy is important as with an increase in degeneracy there is a greater distribution of molecules in various energy states. Having found a formula for the rotational vibration energy levels and knowing that a transition occurs if there is a transition dipole moment, an experiment can be designed to measure the rotational energy levels.

$$M_{J,J} = \int \psi_J^* \hat{M} \psi_J dq$$

The equation above represents how for a dipole moment is directly related to the azimuthal quantum number.

There is a great importance on what transitions need to occur from these vibrations. Now we know that they occur and what degeneracy they have it is important that when the molecule is irradiated the light responds in a certain way. For the rotational spectrum to show there needs to be a permanent dipole moment present. So the molecular quantum number has to move for a transition to be present this is represented by the following:

$$M_{J'J} \neq 0 \; if \; J' = J \pm 1$$

$$M_{J'J} = 0 \; if \; j' \neq J \pm 1$$

This leaves the selection rule:

$$\Delta J = \pm 1.$$

Knowing an increase in J means that there is an increase energy on absorption and a reduction in J means that there has been an emission. This means that the absorption frequencies in a vibration can be given by:

$$hv = \Delta E = E_{j+1} - E_j = Bh(J+1)(J+2) - BhJ(J+1)$$
$$hv = 2BhJ(J+1)$$
$$v = 2B(J+1).$$

The spectrum therefore shows absorption lines at 2B, 4B etc. because there is only a small splitting energy between these the low J values are mostly filled at room temperature. This is governed by the Boltzmann factor.

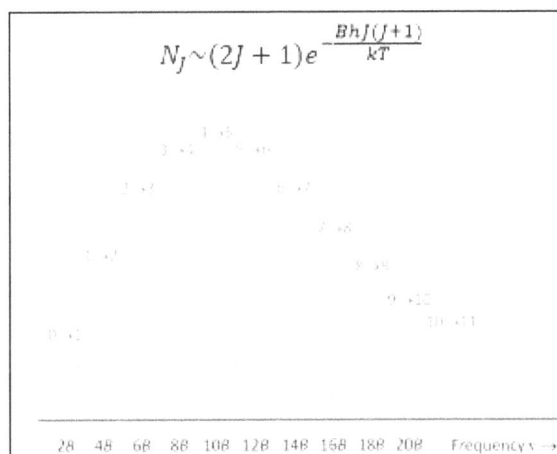

The spacing in the diagram can give the value of B. This is a great example of how rotational spectroscopy can then be used to find a large number of key properties of the molecule being analysed.

Now that we have B and the inertia value the bond length can be calculated. This is useful for analysis methods and for further calculation:

$$I = m_r r^2$$

In a polyatomic molecule it can be much harder to find bond lengths of each of the molecules. This can be helped by using isotopic substitution and assuming the bond lengths are unchanged on isotopic substitution which is not quite true. The observed frequencies work well for low J values but not for high J values. When centrifugal stretching of the molecule is taken into account a correction term needs to be added. This shows how the centrifugal constant has quite a large effect on the matter.

$$E = BhJ(J+1) - DhJ^2(J+1)^2$$

This is where D is the centrifugal distortion constant, this is inversely proportional to the moment of inertia.

Non-linear Polyatomic Moleculeis

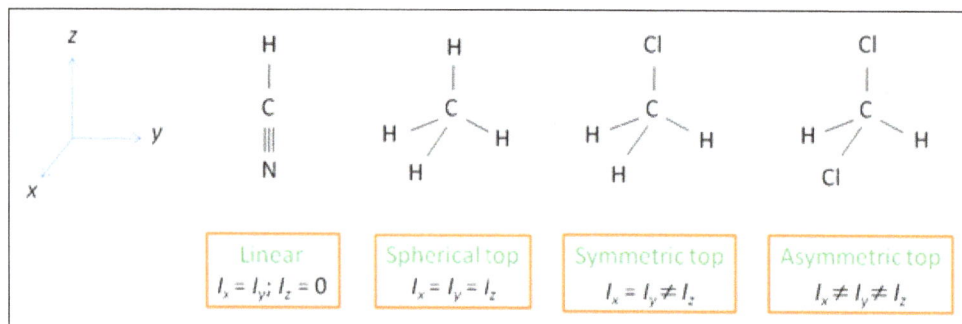

Linear	Spherical top	Symmetric top	Asymmetric top
$I_x = I_y; I_z = 0$	$I_x = I_y = I_z$	$I_x = I_y \neq I_z$	$I_x \neq I_y \neq I_z$

Spherical top molecules have similar rotational levels to linear molecules but they do not have a dipole moment. This means there cannot be a rotation spectra. Symmetric and asymmetric top molecules do give vibration spectra. Knowing the moment of inertia of these molecules allows the prediction of bond lengths and bond angles. On the whole, rotational spectroscopy provides a useful method of determining molecular structure.

The rotational spectroscopy of molecules typically (but by no means always) occurs in the microwave region of the spectrum. In solids, molecules are usually not free to rotate, and in liquids collisions normally render absorption featureless; we therefore consider only the rotational spectroscopy of gaseous molecules.

Rotational spectroscopy measures a high-resolution spectrum where the spectral pattern is determined by the three-dimensional structure of the molecule. The quantized energy levels for the spectroscopy come from the overall rotational motion of the molecule. The rotational kinetic energy is determined by the three moments-of-inertia in the principal axis system. Any changes in the mass distribution will produce a different energy level structure and spectroscopic transition frequencies. Therefore, structural isomers have distinct rotational spectra but enantiomers, which have the same set of bond lengths and bond angles, have identical rotational spectra. In order for the rotational motion of the molecule to couple with light, it is necessary for the molecule to have a permanent dipole moment. Although there are special cases of chiral molecules with elements of molecular symmetry, they generally have C1-symmetry and are polar.

Standard rotational spectroscopy can often distinguish well between different isomers, provided they are not enantiomers. Chiral rotational spectroscopy can distinguish well between different isomers including enantiomers. It may find particular use, therefore, in the analysis of molecules with multiple chiral centers, which permit an exponentially large number of different stereoisomers, many of which are enantiomers. This could see chiral rotational spectroscopy find particular use in the food and pharmaceutical industries, where different isomers must be individually justified and molecules with multiple chiral centers are recognized as being "challenging".

Consider a sample of tartaric acid. The $N = 2$ chiral centers permit $2N = 4$ stereoisomers, two of which are equivalent, leaving three distinct stereoisomers. One of these, mesotartaric acid, is achiral, whereas the other two, l-tartaric acid and d-tartaric acid, are enantiomers. Depicted in figure is the $2_{-2} \leftarrow 1_{-1}$ rotational line for a 50:n:(50 − n) mixture of mesotartaric acid, l-tartaric acid, and d-tartaric acid in the absence of light, where n is *any* number between 0 and 50 inclusive and we have assumed equal line strengths for each molecule. The contribution due to mesotartaric acid appears well separated from that due to l-tartaric acid and d-tartaric acid. The spectrum gives no information, however, about the relative abundances of l-tartaric acid and d-tartaric acid (determined by n), only their combination ($n + (50 - n) = 50$, independent of n). Depicted in figure is the $2_{-2,0} \leftarrow 1_{-1,\pm 1}$ rotational line for a 50:20:30 mixture in the presence of light with $I = 1.0000 \times 10^{12}\, kg\, s^{-3}$, $2\pi / |k| = 5.320 \times 10^{-7}\, m$, and $\sigma = 1$. Contributions due to all *three* stereoisomers now appear well distinguished while yielding a wealth of new information, as claimed.

Rotational spectra are sufficiently sparse that the analysis of molecules with significantly more chiral centers in this way should not be met with any fundamental difficulties. This ability to distinguish well and in a chirally sensitive manner between subtly different molecular forms persists moreover for more general mixtures containing multiple types of molecule. The chirally sensitive analysis of complicated mixtures using traditional techniques represents a serious challenge. Indeed, it was suggested in 2014 that "only one mixture analysis (based upon circular dichroism, vibrational circular dichroism or Raman optical activity) was reported so far", although the use of chiral microwave three wave mixing to analyze various mixtures has now been well demonstrated.

Complementary Techniques

High-resolution rotational spectroscopy, and most often the pure rotational spectra (microwave spectroscopy), is a complementary source of structural information for gaseous molecules in an electron diffraction structure analysis. Its geometrical findings usually appear in terms of operational parameters r_o and r_s. The r_o internuclear distance is obtained from the rotational constants and usually, though not always, refers to the ground vibrational state. It depends strongly on the isotopic composition and may considerably differ from the equilibrium distance. The r_s internuclear distance is determined from the isotopic substitution coordinates of the respective atoms. It depends only slightly on the isotopic composition and thus may only slightly differ from the equilibrium internuclear distance. However, neither r_o nor r_s have rigorously defined physical meaning. Vibrational corrections may also be applied to the primary data from rotational spectroscopy, and consequently, internuclear distances between average nuclear positions may be obtained; they are labeled r_z and correspond to the raodistances from electron diffraction. They are very useful in the characterization of bond angles, but are less useful for describing bond lengths because they are averages projected onto the direction of the lines connecting equilibrium nuclear positions. An excellent representation of bond length is the r_g distance, as it is a real distance averaged over all molecular vibrations. The most unambiguous representation of molecular geometry is in terms of the r_e equilibrium distances. The uncertainties in the determination of r_e parameters no longer refer merely to precision, but can be considered to characterize the accuracy of the structure determination.

If the electron diffraction structural parameters are converted into r_α^0 distances and the rotational constants from the microwave spectra (A_o, B_o, and C_o) into A_z, B_z, and C_z rotational constants, referring to the distances between average nuclear positions in the ground vibrational state, then the measurements from the two techniques can be used in a combined way in the structure refinement. The conversion can be carried out by means of a harmonic force field and normal-coordinate analysis. Note that although the conversion is done on the geometrical parameters for electron diffraction, it is done on the rotational constants for microwave spectroscopy. In a combined refinement, the quantity to be minimized may be, for example,

$$\sum_{k=1}^{n}[s_k M_k^E (s_k) - s_k M_k^T(s_k)] \xleftarrow{\ 2\ } W_k + \sum_{j=a,b,c} (I_j^E - I_j^T)^2 \leftarrow W_j$$

where M^E and M^T represent the experimental and calculated molecular intensities and I^E and I^T the experimental and calculated principal moments of inertia, respectively. The latter are calculated from the respective rotational constants. The choice of the relative weights W_k and W_j is a delicate matter for which no general formula is available. Depending on the data and the structural problem under investigation, various schemes have to be tested. Whereas the electron diffraction intensity measurements may number over a hundred, the introduction of three, or sometimes even fewer, pieces of data from high-resolution spectroscopy may eliminate the ambiguity of a structure determination and result in noticeable enhancement of accuracy. This is because the pattern of correlation among the parameters may be strikingly different for the two sets of observables. Thus, for example, for sulfone molecules, such as SO_2Cl_2 and SO_2F_2, the tetrahedral sulfur bond configuration results in rather closely spaced nonbonded distances, which hinder their precise determination from electron diffraction, whereas their bond lengths can be determined with high precision. However, the distances O–O as well as Cl–Cl for SO_2Cl_2 or F–F for SO_2F_2 can be determined directly from the principal moments of inertia with no similar information on the bond lengths of these molecules from the microwave spectra. Even without vibrational corrections, the utilization of complementary information from the two techniques has great importance. In such a case, however, the final refinements should be done by removing data from the other technique lest systematic errors be introduced by the uncorrected measurements.

The above-discussed example illustrates the point that a general scheme for combined application of different techniques could hardly be useful. Indeed, a very broad range of approaches may be followed when incorporating additional information in an electron diffraction structure analysis. The procedure may range from the use of some chemical experience in constructing models in the initial stages of the analysis to built-in calculated vibrational amplitudes from normal-coordinate analysis in the least-squares refinements to assuming differences of similar bond lengths or relative contributions of different conformersfrom other techniques. Quantum chemical calculations have played an increasing role in electron diffraction structure analysis in many ways. They can provide computed parameter differences for the least-squares refinement in case of closely spaced distances, vibrational frequencies, and force fields for normal-coordinate analysis to produce vibrational amplitudes, information on possible molecular models that correspond to the global minimum and low-energy local minima on the potential energy surface together with their energy differences and energy barriers, types and relative amounts of different conformers possibly present in the vapor based on their energy differences, and so on.

Vibrational frequencies, either from infrared or Raman spectroscopic measurements or from computation, may also be incorporated into the structure refinement; this has been especially useful for small molecules with large-amplitude vibrations. Besides other gas-phase experimental methods, liquid crystal NMR has also proven useful to combine with electron diffraction data. The inclusion of dipolar couplingsfrom solution-phase NMR studies in the analysis increases the precision of the structure determination as has been shown for different benzene derivatives. Mass spectrometry has been very helpful in providing knowledge about the vapor composition and in the optimization of the experimental conditions. In case of complex vapor composition, computations are also an invaluable aid to the electron diffraction analysis. An example is chromiumdichloride that has different oligomers in its vapor besides the monomer that, according to the computations, has a C_{2v}-symmetry structure due to the Renner–Teller effect. Utilizing the information from computations, the structure of the monomer as well as the oligomers could be determined from electron diffraction; without this additional information, this proved to be impossible. Molecules with different conformers present in their vapors can also be better studied with the approach of joint analysis and thus more parameters can be determined. Thus, enhanced amount of structural information for conformers include their relative amounts and their temperature dependence, energy barriers, and the subtle geometrical differences between their structures.

Comparison with structural information from X-ray crystallography has also been instrumental. In detailed comparison, it has to be born in mind that X-ray diffraction yields distances between the centroids of the electron density distribution rather than internuclear distances, but they can well approximate the distances between average nuclear positions. Thus, the best comparison may be in terms of the r_{α}^{0} structures. If the molecular structures of the same substance determined by gas-phase electron diffraction and X-ray crystallography show significant differences (i.e., beyond experimental uncertainties), those differences can be ascribed to the impact of intermolecular interactions in the crystal.

Today, by far, most gas-phase electron diffraction analyses are carried out in combination with other techniques. The most often applied additional method is quantum chemical calculations. Often, data from rotational and vibrational spectroscopy and from liquid crystal NMR are also incorporated in the joint analysis in order to get a deeper and more reliable insight into the structure of the molecule. Different structural techniques usually are sensitive to different characteristics of a molecule and may thus ideally complement each other, and this is what makes the joint analysis valuable and the way of the future.

Microwave Rotational Spectroscopy

Microwave rotational spectroscopy uses microwave radiation to measure the energies of rotational transitions for molecules in the gas phase. It accomplishes this through the interaction of the electric dipole moment of the molecules with the electromagnetic field of the exciting microwave photon.

To probe the pure rotational transitions for molecules, scientists use microwave rotational spectroscopy. This spectroscopy utilizes photons in the microwave range to cause transitions between the quantum rotational energy levels of a gas molecule. The reason why the sample must be in the gas phase is due to intermolecular interactions hindering rotations in the liquid and solid phases of the molecule. For microwave spectroscopy, molecules can be broken down into 5 categories based on their shape and the inertia around their 3 orthogonal rotational axes. These 5 categories include diatomic molecules, linear molecules, spherical tops, symmetric tops and asymmetric tops.

Classical Mechanics

The Hamiltonian solution to the rigid rotor is,

$$H = T$$

since,

$$H = T + V$$

where T is kinetic energy and V is potential energy. Potential energy, V, is 0 because there is no resistance to the rotation (similar to a particle in a box model).

Since $H = T$, we can also say that:

$$T = \frac{1}{2} \sum m_i v_i^2.$$

However, we have to determine v_i in terms of rotation since we are dealing with rotation. Since,

$$\omega = \frac{v}{r}$$

Where ω = angular velocity, we can say that:

$$v_i = \omega X r_i$$

Thus we can rewrite the T equation as:

$$T = \frac{1}{2} \sum m_i v_i (\omega X r_i)$$

Since ω is a scalar constant, we can rewrite the T equation as:

$$T = \frac{\omega}{2} \sum m_i (v_i X r_i) = \frac{\omega}{2} \sum l_i = \omega \frac{L}{2}$$

where l_i is the angular momentum of the *ith* particle, and L is the angular momentum of the entire system. Also, we know from physics that,

$$L = I\omega$$

where I is the moment of inertia of the rigid body relative to the axis of rotation. We can rewrite the T equation as,

$$T = \omega \frac{I\omega}{2} = \frac{1}{2} I\omega^2.$$

Quantum Mechanics

The internal Hamiltonian, H, is:

$$H = \frac{i^2\hbar^2}{2I}$$

and the Schrödinger Equation for rigid rotor is:

$$\frac{i^2\hbar^2}{2I}\psi = E\psi$$

Thus, we get:

$$E_n = \frac{J(J+1)h^2}{8\pi^2 I}$$

where J is a rotational quantum number and $\hbar$ is the reduced Planck's constant. However, if we let:

$$B = \frac{h}{8\pi^2 I}$$

where B is a rotational constant, then we can substitute it into the E_n equation and get:

$$E_n = J(J+1)Bh$$

Considering the transition energy between two energy levels, the difference is a multiple of 2. That is, from $J = 0$ to $J = 1$, the $\Delta E_{0\to 1}$ is 2Bh and from $J = 1$ to $J = 2$, the $\Delta E_{1\to 2}$ is 4Bh.

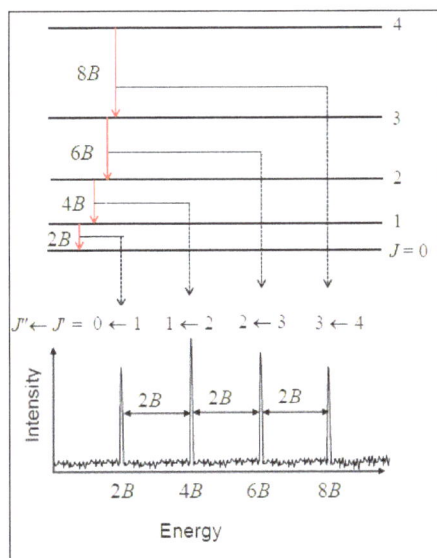

Figure: Energy levels and line positions calculated in the rigid rotor approximation. This diagram illustrates how transitions between the rotational energy levels of molecules map onto the energies at which these transitions are observed during laboratory experiments.

Theory

When a gas molecule is irradiated with microwave radiation, a photon can be absorbed through the interaction of the photon's electronic field with the electrons in the molecules. For the microwave region this energy absorption is in the range needed to cause transitions between rotational states of the molecule. However, only molecules with a permanent dipole that changes upon rotation can be investigated using microwave spectroscopy. This is due to the fact that their must be a charge difference across the molecule for the oscillating electric field of the photon to impart a torque upon the molecule around an axis that is perpendicular to this dipole and that passes through the molecules center of mass.

This interaction can be expressed by the transition dipole moment for the transition between two rotational states,

$$\text{Probability of Transition} = \int \psi_{rot}(F)\hat{\mu}\psi_{rot}(I)d\tau$$

where $\Psi_{rot(F)}$ is the complex conjugate of the wave function for the final rotational state, $\Psi_{rot(I)}$ is the wave function of the initial rotational state and μ is the dipole moment operator with Cartesian coordinates of μ_x, μ_y, μ_z. For this integral to be nonzero the integrand must be an even function. This is due to the fact that any odd function integrated from negative infinity to positive infinity, or any other symmetric limits, is always zero.

In addition to the constraints imposed by the transition moment integral, transitions between rotational states are also limited by the nature of the photon itself. A photon contains one unit of angular momentum, so when it interacts with a molecule it can only impart one unit of angular momentum to the molecule. This leads to the selection rule that a transition can only occur between rotational energy levels that are only one quantum rotation level (J) away from another.

$$\Delta J = \pm 1$$

The transition moment integral and the selection rule for rotational transitions tell if a transition from one rotational state to another is allowed. However, what these do not take into account is whether or not the state being transitioned from is actually populated, meaning that the molecule is in that energy state. This leads to the concept of the Boltzmann distribution of states. The Boltzmann distribution is a statistical distribution of energy states for an ensemble of molecules based on the temperature of the sample.

$$\frac{n_J}{n_0} = \frac{e^{(-E_{rot}(J)/RT)}}{\sum_{J=1}^{J=n} e^{(-E_{rot}(J)/RT)}}$$

where $E_{rot(J)}$ is the molar energy of the J rotational energy state of the molecule,

- R is the gas constant,
- T is the temperature of the sample,
- n(J) is the number of molecules in the J rotational level, and
- n_0 is the total number of molecules in the sample.

This distribution of energy states is the main contributing factor for the observed absorption intensity distributions seen in the microwave spectrum. This distribution makes it so that the absorption peaks that correspond to the transition from the energy state with the largest population based on the Boltzmann equation will have the largest absorption peak, with the peaks on either side steadily decreasing.

Degrees of Freedom

A molecule can have three types of degrees of freedom and a total of 3N degrees of freedom, where N equals the number of atoms in the molecule. These degrees of freedom can be broken down into 3 categories.

1. Translational: These are the simplest of the degrees of freedom. These entail the movement of the entire molecule's center of mass. This movement can be completely described by three orthogonal vectors and thus contains 3 degrees of freedom.

2. Rotational: These are rotations around the center of mass of the molecule and like the translational movement they can be completely described by three orthogonal vectors. This again means that this category contains only 3 degrees of freedom. However, in the case of a linear molecule only two degrees of freedom are present due to the rotation along the bonds in the molecule having a negligible inertia.

3. Vibrational: These are any other types of movement not assigned to rotational or translational movement and thus there are 3N − 6 degrees of vibrational freedom for a nonlinear molecule and 3N − 5 for a linear molecule. These vibrations include bending, stretching, wagging and many other aptly named internal movements of a molecule. These various vibrations arise due to the numerous combinations of different stretches, contractions, and bends that can occur between the bonds of atoms in the molecule.

Each of these degrees of freedom is able to store energy. However, In the case of rotational and vibrational degrees of freedom, energy can only be stored in discrete amounts. This is due to the quantized break down of energy levels in a molecule described by quantum mechanics. In the case of rotations the energy stored is dependent on the rotational inertia of the gas along with the corresponding quantum number describing the energy level.

Rotational Symmetries

To analyze molecules for rotational spectroscopy, we can break molecules down into 5 categories based on their shapes and their moments of inertia around their 3 orthogonal rotational axes:

1. Diatomic Molecules

2. Linear Molecules

3. Spherical Tops

4. Symmetrical Tops

5. Asymmetrical Tops.

Diatomic Molecules

The rotations of a diatomic molecule can be modeled as a rigid rotor. This rigid rotor model has two masses attached to each other with a fixed distance between the two masses.

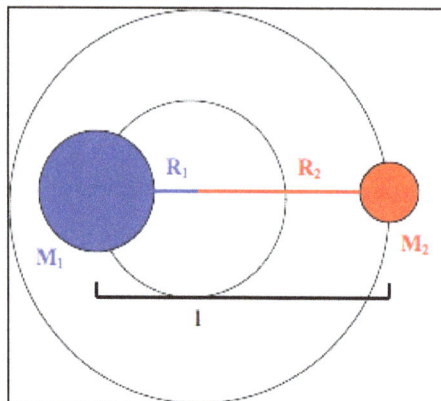

It has an inertia (I) that is equal to the square of the fixed distance between the two masses multiplied by the reduced mass of the rigid rotor.

$$I_e = \mu r_e^2$$

$$\mu = \frac{m_1 m_2}{m_1 + m_2}$$

Using quantum mechanical calculations it can be shown that the energy levels of the rigid rotator depend on the inertia of the rigid rotator and the quantum rotational number J.

$$E(J) = B_e J(J+1)$$

$$B_e = \frac{h}{8\pi^2 c I_e}$$

However, this rigid rotor model fails to take into account that bonds do not act like a rod with a fixed distance, but like a spring. This means that as the angular velocity of the molecule increases so does the distance between the atoms. This leads us to the nonrigid rotor model in which a centrifugal distortion term (D_e) is added to the energy equation to account for this stretching during rotation.

$$E(J)(cm^{-1}) = B_e J(J+1) - D_e J^2(J+1)^2$$

This means that for a diatomic molecule the transitional energy between two rotational states equals,

$$E = B_e[J'(J'+1) - J''(J''+1)] - D_e[J'^2(J'+1)^2 - J''^2(J'+1)^2]$$

Where J' is the quantum number of the final rotational energy state and J" is the quantum number of the initial rotational energy state. Using the selection rule of $\Delta J = \pm 1$ the spacing between peaks in the microwave absorption spectrum of a diatomic molecule will equal.

$$E_R = (2B_e - 4D_e) + (2B_e - 12D_e)J'' - 4D_eJ''^3$$

Linear Molecules

Linear molecules behave in the same way as diatomic molecules when it comes to rotations. For this reason they can be modeled as a non-rigid rotor just like diatomic molecules. This means that linear molecule have the same equation for their rotational energy levels. The only difference is there are now more masses along the rotor. This means that the inertia is now the sum of the distance between each mass and the center of mass of the rotor multiplied by the square of the distance between them.

$$I_e = \sum_{j=1}^{n} m_j r_{ej}^2$$

Where m_j is the mass of the jth mass on the rotor and r_{ej} is the equilibrium distance between the jth mass and the center of mass of the rotor.

Spherical Tops

Spherical tops are molecules in which all three orthogonal rotations have equal inertia and they are highly symmetrical. This means that the molecule has no dipole and for this reason spherical tops do not give a microwave rotational spectrum.

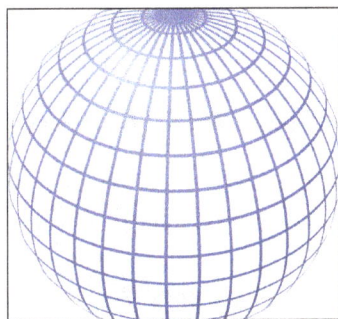

Figure: Geometrical example of a spherical top.

Examples:

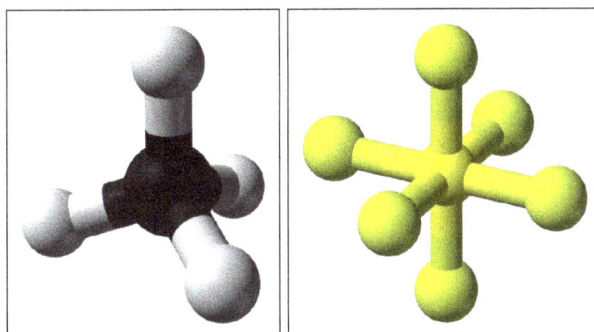

Symmetrical Tops

Symmetrical tops are molecules with two rotational axes that have the same inertia and one unique rotational axis with a different inertia. Symmetrical tops can be divided into two categories based on the relationship between the inertia of the unique axis and the inertia of the two axes with equivalent inertia. If the unique rotational axis has a greater inertia than the degenerate axes the molecule is called an oblate symmetrical top. If the unique rotational axis has a lower inertia than the degenerate axes the molecule is called a prolate symmetrical top. For simplification think of these two categories as either frisbees for oblate tops or footballs for prolate tops.

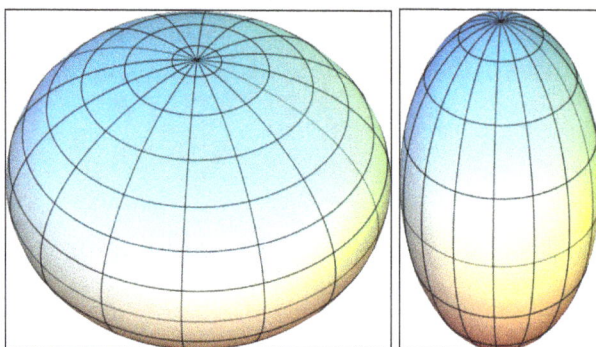

Figure: Symmetric Tops: (Left) Geometrical example of an oblate top and (right) a prolate top.

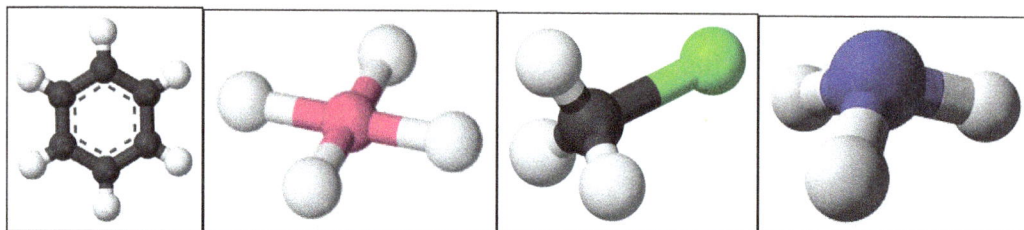

Figure: Examples of symmetric tops. Benzene (oblate) XeF_4 (oblate) $ClCH_3$ (prolate) NH_3 (prolate).

In the case of linear molecules there is one degenerate rotational axis which in turn has a single rotational constant. With symmetrical tops now there is one unique axis and two degenerate axes. This means an additional rotational constant is needed to describe the energy levels of a symmetrical top. In addition to the rotational constant an additional quantum number must be introduced to describe the rotational energy levels of the symmetric top. These two additions give us the following rotational energy levels of a prolate and oblate symmetric top.

$$E(J,K)(cm^{-1}) = Be * J(J+1) + (A_e - B_e)K^2$$

Where B_e is the rotational constant of the unique axis, A_e is the rotational constant of the degenerate axes, J is the total rotational angular momentum quantum number and K is the quantum number that represents the portion of the total angular momentum that lies along the unique rotational axis. This leads to the property that K is always equal to or less than J. Thus we get the two selection rules for symmetric tops.

$$\Delta J = 0, \pm 1$$

$$\Delta K = 0$$

when, $K \neq 0$

$$\Delta J = \pm 1$$

$$\Delta K = 0$$

when, $K = 0$

However, like the rigid rotor approximation for linear molecules, we must also take into account the elasticity of the bonds in symmetric tops. Therefore, in a similar manner to the rigid rotor we add a centrifugal coupling term, but this time we have one for each quantum number and one for the coupling between the two.

$$E_{(J,K)}(cm^{-1}) = B_e J(J+1) - D_{eJ} J^2 (J+1)^2 + (A_e - B_e) * K^2$$

$$- D_{ek} K^4 - D_{ejk} J(J+1) K^2$$

Asymmetrical Tops

Asymmetrical tops have three orthogonal rotational axes that all have different moments of inertia and most molecules fall into this category. Unlike linear molecules and symmetric tops these types of molecules do not have a simplified energy equation to determine the energy levels of the rotations. These types of molecules do not follow a specific pattern and usually have very complex microwave spectra.

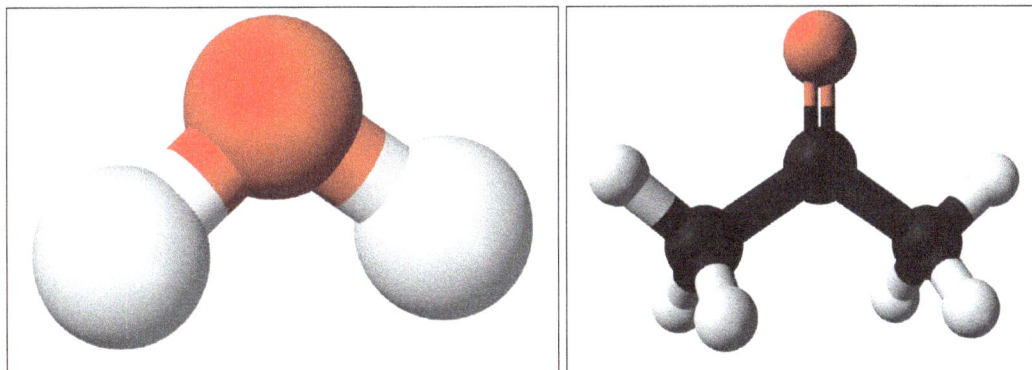

Figure: Molecular examples of asymmetrical tops. (left) water and (right) acetone.

Additional Rotationally Sensitive Spectroscopies

In addition to microwave spectroscopy, IR spectroscopy can also be used to probe rotational transitions in a molecule. However, in the case of IR spectroscopy the rotational transitions are coupled to the vibrational transitions of the molecule. One other spectroscopy that can probe the rotational transitions in a molecule is Raman spectroscopy, which uses UV-visible light scattering to determine energy levels in a molecule. However, a very high sensitivity detector must be used to analyze rotational energy levels of a molecule.

Rotational Spectroscopy of Diatomic Molecules

The rotation of a diatomic molecule can be described by the rigid rotor model. To imagine this model think of a spinning dumbbell. The dumbbell has two masses set at a fixed distance from one another and spins around its center of mass (COM). This model can be further simplified using the concept of reduced mass which allows the problem to be treated as a single body system.

Similar to most quantum mechanical systems our model can be completely described by its wave function. Therefore, when we attempt to solve for the energy we are lead to the Schrödinger Equation. In the context of the rigid rotor where there is a natural center (rotation around the COM) the wave functions are best described in spherical coordinates. In addition to having pure rotational spectra diatomic molecules have rotational spectra associated with their vibrational spectra. The order of magnitude differs greatly between the two with the rotational transitions having energy proportional to 1-10 cm^{-1} (microwave radiation) and the vibrational transitions having energy proportional to 100-3,000 cm^{-1} (infrared radiation). Rotational spectroscopy is therefore referred to as microwave spectroscopy.

Rigid Rotor Model

A diatomic molecule consists of two masses bound together. The distance between the masses, or the bond length, (l) can be considered fixed because the level of vibration in the bond is small compared to the bond length. As the molecule rotates it does so around its COM with a frequency of rotation of vrotvrot given in radians per second.

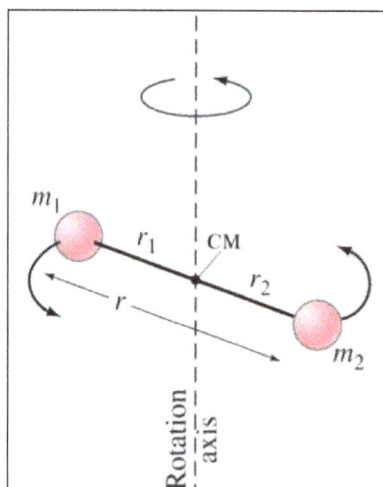

Figure: Rigid Rotor Model of a Diatomic Molecule.

Reduced Mass

The system can be simplified using the concept of reduced mass which allows it to be treated as one rotating body. The system can be entirely described by the fixed distance between the two masses instead of their individual radii of rotation. Relationships between the radii of rotation and bond length are derived from the COM given by:

$$M_1R_1 = M_2R_2,$$

where l is the sum of the two radii of rotation:

$$l = R_1 + R_2.$$

Through simple algebra both radii can be found in terms of their masses and bond length:

$$R_1 = \frac{M_2}{M_1 + M_2} l$$

and

$$R_2 = \frac{M_1}{M_1 + M_2} l.$$

The kinetic energy of the system, T, is sum of the kinetic energy for each mass:

$$T = \frac{M_1 v_1^2 + M_2 v_2^2}{2},$$

where,

$$v_1 = 2\pi R_1 v_{rot}$$

and,

$$v_2 = 2\pi R_2 v_{rot}.$$

Using the angular velocity,

$$\omega = 2\pi v_{rot}$$

the kinetic energy can now be written as:

$$T = \frac{M_1 R_1^2 + M_2 R_2^2}{2} \omega.$$

With the moment of inertia,

$$I = M_1 R_1^2 + M_2 R_2^2,$$

the kinetic energy can be further simplified:

$$T = \frac{I\omega^2}{2}.$$

The moment of inertia can be rewritten by plugging in for R_1 and R_2:

$$I = \frac{M_1 M_2}{M_1 + M_2} l^2,$$

Where,

$$\frac{M_1 M_2}{M_1 + M_2}$$

is the reduced mass, μ. The moment of inertia and the system are now solely defined by a single mass, μ, and a single length, l:

$$I = \mu l^2.$$

Angular Momentum

Another important concept when dealing with rotating systems is the the angular momentum defined by,

$$L = I\omega.$$

Looking back at the kinetic energy:

$$T = \frac{I\omega^2}{2} = \frac{I^2\omega^2}{2I} = \frac{L^2}{2I}.$$

The angular momentum can now be described in terms of the moment of inertia and kinetic energy: $L^2 = 2IT$.

Setting up the Schrödinger Equation

The wave functions for the rigid rotor model are found from solving the time-independent Schrödinger Equation:

$$\hat{H}\psi = E\psi.$$

Where the Hamiltonian Operator is:

$$\hat{H} - \frac{-\hbar}{2\mu}\nabla^2 + V(r)$$

where ∇^2 is the Laplacian Operator and can be expressed in either Cartesian coordinates:

$$\nabla^2 = \frac{\partial^2}{\partial x^2} + \frac{\partial^2}{\partial y^2} + \frac{\partial^2}{\partial z^2}$$

or in spherical coordinates:

$$\nabla^2 = \frac{1}{r^2}\frac{\partial}{\partial r}\left(r^2\frac{\partial}{\partial r}\right) + \frac{1}{r^2\sin\theta}\frac{\partial}{\partial\theta}\left(\sin\theta\frac{\partial}{\partial\theta}\right) + \frac{1}{r^2\sin^2\theta}\frac{\partial^2}{\partial\phi^2}.$$

At this point it is important to incorporate two assumptions:

- The distance between the two masses is fixed. This causes the terms in the Laplacian containing $\dfrac{\partial}{\partial r}$ to be zero.

- The orientation of the masses is completely described by θ and ϕ and in the absence of electric or magnetic fields the energy is independent of orientation. This causes the potential energy portion of the Hamiltonian to be zero.

The wave functions $\psi(\theta,\phi)$ are customarily represented by $Y(\theta,\phi)$ and are called spherical harmonics.

The Hamiltonian Operator can now be written:

$$\hat{H} = \hat{T} = \frac{-\hbar^2}{2\mu l^2}\left[\frac{1}{\sin\theta}\frac{\partial}{\partial\theta}\left(\sin\theta\frac{\partial}{\partial\theta}\right) + \frac{1}{\sin\theta}\frac{\partial^2}{\partial\phi^2}\right]$$

with the Angular Momentum Operator being defined:

$$\hat{L} = 2I\hat{T}$$

$$\hat{L} = -\hbar^2\left[\frac{1}{\sin\theta}\frac{\partial}{\partial\theta}\left(\sin\theta\frac{\partial}{\partial\theta}\right) + \frac{1}{\sin\theta}\frac{\partial^2}{\partial\phi^2}\right]$$

The Schrödinger Equation now expressed:

$$\frac{-\hbar^2}{2I}\left[\frac{1}{\sin\theta}\frac{\partial}{\partial\theta}\left(\sin\theta\frac{\partial}{\partial\theta}\right) + \frac{1}{\sin\theta}\frac{\partial^2}{\partial\phi^2}\right]Y(\theta,\phi) = EY(\theta,\phi).$$

Solving the Schrödinger Equation

The Schrödinger Equation can be solved using separation of variables.

Step 1:

Let $Y(\theta,\phi) = \Theta(\theta)\Phi(\phi)$ and substitute: $\beta = \dfrac{2IE}{\hbar^2}$.

Set the Schrödinger Equation equal to zero:

$$\frac{\sin\theta}{\Theta(\theta)}\frac{d}{d\theta}\left(\sin\theta\frac{d\Theta}{d\theta}\right) + \beta\sin^2\theta + \frac{1}{\Phi(\phi)}\frac{d^2\Phi}{d\phi^2} = 0$$

Step 2: Because the terms containing $\Theta(\theta)$ are equal to the terms containing $\Phi(\phi)$ they must equal the same constant in order to be defined for all values:

$$\frac{\sin\theta}{\Theta(\theta)}\frac{d}{d\theta}\left(\sin\theta\frac{d\Theta}{d\theta}\right) + \beta\sin^2\theta = m^2$$

$$\frac{1}{\Phi(\phi)}\frac{d^2\Phi}{d\phi^2} = -m^2$$

Step 3: Solving for Φ is fairly simple and yields:

$$\Phi(\phi) = \frac{1}{\sqrt{2\pi}}e^{im\phi}$$

where $m = 0, \pm1, \pm2, \ldots$

Solving for θ is considerably more complicated but gives the quantized result:

$$\beta = J(J+1)$$

where J is the rotational level with J=0,1,2,...

Step 4: The energy is quantized by expressing in terms of β:

$$E = \frac{\hbar^2\beta}{2I}$$

Step 5: Using the rotational constant, $B = \dfrac{\hbar^2}{2I}$, the energy is further simplified: $E = BJ(J+1)$.

Energy of Rotational Transitions

When a molecule is irradiated with photons of light it may absorb the radiation and undergo an energy transition. The energy of the transition must be equivalent to the energy of the photon of light absorbed given by: $E = h\nu$. For a diatomic molecule the energy difference between rotational levels (J to J+1) is given by:

$$E_{J+1} - E_J = B(J+1)(J+2) - BJ(J=1) = 2B(J+1)$$

with J=0, 1, 2,...

Because the difference of energy between rotational levels is in the microwave region (1-10 cm^{-1}) rotational spectroscopy is commonly called microwave spectroscopy. In spectroscopy it is customary to represent energy in wave numbers (cm^{-1}), in this notation B is written as $\tilde{B}$. To convert from units of energy to wave numbers simply divide by h and c, where c is the speed of light in cm/s (c=2.998e10 cm/s). In wave numbers $\tilde{B} = \dfrac{h}{8\pi cI}$.

Predicts the rotational spectra of a diatomic molecule to have several peaks spaced by $2\tilde{B}$. This contrasts vibrational spectra which have only one fundamental peak for each vibrational mode. From the rotational spectrum of a diatomic molecule the bond length can be determined. Because $\tilde{B}$ is a function of I and therefore a function of I (bond length), so l can be readily solved for:

$$l = \sqrt{\frac{h}{8\pi^2 c\tilde{B}\mu}}.$$

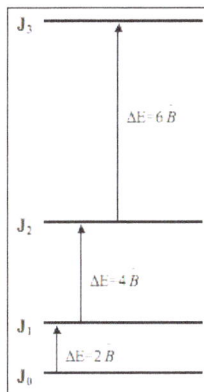

Figure: Rotational Energy Levels.

Centrifugal Distortion

As molecules are excited to higher rotational energies they spin at a faster rate. The faster rate of spin increases the centrifugal force pushing outward on the molecules resulting in a longer average bond length. Looking back, B and l are inversely related. Therefore the addition of centrifugal distortion at higher rotational levels decreases the spacing between rotational levels. The correction for the centrifugal distortion may be found through perturbation theory:

$$E_J = \tilde{B}J(J+1) - \tilde{D}J^2(J+1)^2.$$

Rotation-vibration Transitions

Rotational transitions are on the order of 1-10 cm^{-1}, while vibrational transitions are on the order of 1000 cm^{-1}. The difference of magnitude between the energy transitions allow rotational levels to be superimposed within vibrational levels.

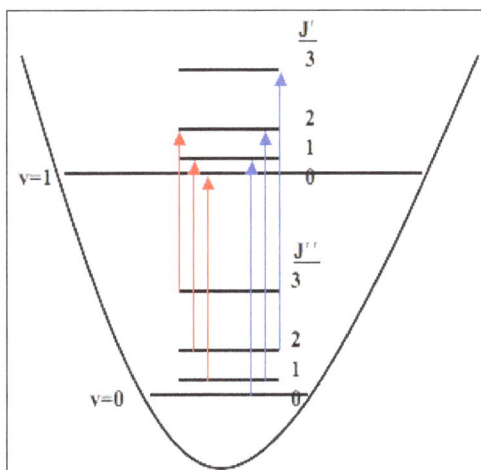

Figure: Rotation-Vibration Transitions.

Combining the energy of the rotational levels $\tilde{E}_J = \tilde{B}J(J+1)$ with the vibrational levels, $\tilde{E}_v = \tilde{\omega}(v+1/2)$, yields the total energy of the respective rotation-vibration levels:

$$\tilde{E}_{v,J} = \tilde{\omega}(v+1/2) + \tilde{B}J(J+1).$$

Following the selection rule, $\Delta J = J \pm 1$, figure shows all of the allowed transitions for the first three rotational states, where J" is the initial state and J' is the final state.

When the $\Delta J = +1$ transitions are considered (blue transitions) the initial energy is given by: $\tilde{E}_{0,J} = \tilde{\omega}(1/2) + \tilde{B}J(J+1)$ and the final energy is given by: $\tilde{E}_{v,J+1} = \tilde{\omega}(3/2) + \tilde{B}(J+1)(J+2)$.

The energy of the transition, $\Delta\tilde{\nu} = \tilde{E}_{1,J+1} - \tilde{E}_{0,J}$, is therefore:

$$\Delta\tilde{\nu} = \tilde{\omega} + 2\tilde{B}(J+1)$$

where J"=0, 1, 2,...

When the $\Delta J = -1$ transitions are considered (red transitions) the initial energy is given by: $\tilde{E}_{v,J} = \tilde{\omega}(1/2) + \tilde{B}J(J+1)$ and the final energy is given by:

$$\tilde{E}_{v,J-1} = \tilde{\omega}(3/2) + \tilde{B}(J-1)(J).$$

The energy of the transition is therefore: $\Delta\tilde{\nu} = \tilde{\omega} - 2\tilde{B}(J)$ where $J" = 1, 2, 3,...$

The difference in energy between the J+1 transitions and J-1 transitions causes splitting of vibrational spectra into two branches. The J-1 transitions, shown by the red lines in figure, are lower in energy than the pure vibrational transition and form the P-branch. The J+1 transitions, shown by the blue lines in figure are higher in energy than the pure vibrational transition and form the R-branch. Notice that because the $\Delta J = \pm 0$ transition is forbidden there is no spectral line associated with the pure vibrational transition. Therefore there is a gap between the P-branch and R-branch, known as the q branch.

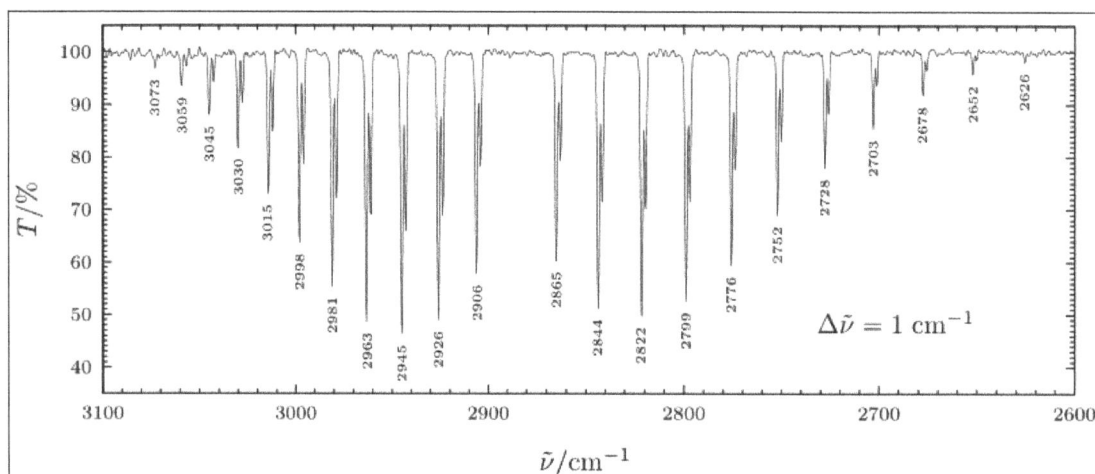

Figure: Rotational-Vibrational Spectrum of HCl.

In the high resolution HCl rotation-vibration spectrum the splitting of the P-branch and R-branch is clearly visible. Due to the small spacing between rotational levels high resolution spectrophotometers are required to distinguish the rotational transitions.

Rotation-vibration Interactions

Recall the Rigid-Rotor assumption that the bond length between two atoms in a diatomic molecule is fixed. However, the anharmonicity correction for the harmonic oscillator predicts the gaps between energy levels to decrease and the equilibrium bond length to increase as higher vibrational levels are accessed. Due to the relationship between the rotational constant and bond length:

$$\tilde{B} = \frac{h}{8\pi^2 c \mu l^2}.$$

The rotational constant is dependent on the vibrational level:

$$\tilde{B}_v = \tilde{B} - \tilde{\alpha}\left(v + \frac{1}{2}\right)$$

Where $\tilde{\alpha}$ is the anharmonicity correction and vv is the vibrational level. As a consequence the spacing between rotational levels decreases at higher vibrational levels and unequal spacing between rotational levels in rotation-vibration spectra occurs.

Including the rotation-vibration interaction the spectra can be predicted.

For the R-branch

$$\tilde{E}_{1,J+1} - \tilde{E}_{0,J}$$

$$\tilde{v} = \left[\tilde{\omega}\left(\frac{3}{2}\right) + \tilde{B}_1\left(J+1\right)\left(J+2\right)\right] - \left[\tilde{\omega}\left(\frac{1}{2}\right) + \tilde{B}_0 J\left(J+1\right)\right]$$

$$\tilde{v} = \tilde{\omega} + \left(\tilde{B}_1 - \tilde{B}_0\right)J^2 + \left(3\tilde{B}_1 - \tilde{B}_0\right)J + 2\tilde{B}_1$$

where J=0, 1, 2,...

For the P-branch

$$\tilde{E}_{1,J-1} - \tilde{E}_{0,J}$$

$$\tilde{v} = \left[\tilde{\omega}\left(\frac{3}{2}\right) + \tilde{B}_1\left(J-1\right)J\right] - \left[\tilde{\omega}\left(\frac{1}{2}\right) + \tilde{B}_0 J\left(J+1\right)\right]$$

$$\tilde{v} = \tilde{\omega} + \left(\tilde{B}_1 - \tilde{B}_0\right)J^2 - \left(\tilde{B}_1 + \tilde{B}_0\right)J$$

where $J = 1, 2, 3,...$

Because $\tilde{B}_1 < \tilde{B}_0$, as J increases:

- Spacing in the R-branch decreases.
- Spacing in the P-branch increases.

Rotation of Polyatomic Molecules

In contrast to diatomic molecules, the rotational motions of polyatomic molecules in three dimensions are characterized by multiple moments of inertia. Typically reflected in an 3×3 inertia tensor. It is common in rigid body mechanics to express in these moments of inertia in *lab-based Cartesian coordinates* via a notation that explicitly identifies the x, y, and z axes such as I_{xx} and I_{xy}, for the components of the inertia tensor.

$$I = \begin{bmatrix} I_{xx} & I_{xy} & I_{xz} \\ I_{yx} & I_{yy} & I_{yz} \\ I_{zx} & I_{zy} & I_{zz} \end{bmatrix}$$

The components of this tensor can be assembled into a matrix given by,

$$I_{xx} = \sum_{k=1}^{N} m_k (y_k^2 + z_k^2)$$

$$I_{yy} = \sum_{k=1}^{N} m_k (x_k^2 + z_k^2)$$

$$I_{zz} = \sum_{k=1}^{N} m_k (x_k^2 + y_k^2)$$

$$I_{yx} = I_{xy} = -\sum_{k=1}^{N} m_k x_k y_k$$

$$I_{zx} = I_{xz} = -\sum_{k=1}^{N} m_k x_k z_k$$

$$I_{zy} = I_{yz} = -\sum_{k=1}^{N} m_k y_k z_k .$$

The rotational motions of polyatomic molecules are characterized by moments of inertia that are defined in a molecule based coordinates with axes that are labeled a, b, and c. Measured in the body frame the inertia matrix ($I = \begin{bmatrix} I_{xx} & I_{xy} & I_{xz} \\ I_{yx} & I_{yy} & I_{yz} \\ I_{zx} & I_{zy} & I_{zz} \end{bmatrix}$) is a constant real symmetric matrix, which can be decomposed into a diagonal matrix, given by,

$$I = \begin{pmatrix} I_a & 0 & 0 \\ 0 & I_b & 0 \\ 0 & 0 & I_c \end{pmatrix}$$

These labels are assigned so that I_c is the largest principal moment of inertia with an order of the three moments set as,

$$I_a < I_b < I_c$$

The rotational kinetic energy operator for a rigid non-linear polyatomic molecule is then expressed as,

$$H_{rot} = \frac{J_a^2}{2I_a} + \frac{J_b^2}{2I_b} + \frac{J_c^2}{2I_c}$$

The components of the quantum mechanical angular momentum operators along the three principal axes are:

$$J_a = -i\hbar \cos\chi \left[\cot\theta \frac{\partial}{\partial\chi} - (\sin\theta)^{-1} \frac{\partial}{\partial\varphi} \right] - -i\hbar \sin\chi \frac{\partial}{\partial\theta}$$

$$J_b = i\hbar \sin\chi \left[\cot\theta \frac{\partial}{\partial\chi} - (\sin\theta)^{-1} \frac{\partial}{\partial\varphi} \right] - -i\hbar \cos\chi \frac{\partial}{\partial\theta}$$

$$J_c = -\frac{i\hbar\partial}{\partial\chi}.$$

The angles θ, φ, and χ are the Euler angles needed to specify the orientation of the rigid molecule relative to a laboratory-fixed coordinate system. The corresponding square of the total angular momentum operator J^2 can be obtained as,

$$J^2 = J_a^2 + J_b^2 + J_c^2$$

$$= -\frac{\partial^2}{\partial\theta^2} - \cot\theta \frac{\partial}{\partial\theta} - \left(\frac{1}{\sin\theta} \right) \left(\frac{\partial^2}{\partial\varphi^2} + \frac{\partial^2}{\partial\chi^2} - 2\cos\theta \frac{\partial^2}{\partial\varphi\partial\chi} \right)$$

and the component along the lab-fixed Z axis is,

$$J_Z = -ih \frac{\partial}{\partial\varphi}.$$

Spherical Tops

When the three principal moment of inertia values are identical, the molecule is termed a spherical top. In this case, the total rotational energy Equation $H_{rot} = \frac{J_a^2}{2I_a} + \frac{J_b^2}{2I_b} + \frac{J_c^2}{2I_c}$ can be expressed in terms of the total angular momentum operator J^2.

$$H_{rot} = \frac{J^2}{2I}$$

As a result, the eigenfunctions of H_{rot} are those of J^2 (and J_a as well as J_z both of which commute with J_2 and with one another; J_z is the component of J along the lab-fixed Z-axis and commutes with J_a because,

$$J_z = -ih\frac{\partial}{\partial\varphi}$$

and,

$$J_a = -ih\frac{\partial}{\partial\chi}$$

act on different angles. The energies associated with such eigenfunctions are,

$$E(J,K,M) = \frac{\hbar^2 J(J+1)}{2I^2}$$

for all K (i.e., J a quantum numbers) ranging from -J to J in unit steps and for all M (i.e., J Z quantum numbers) ranging from -J to J. Each energy level is therefore $(2J+1)^2$ degenarate because there are $2J+1$ possible K values and $2J+1$ possible M values for each J. The eigenfunctions of J^2, J_z and J_a, $|J,M,K\rangle$ are given in terms of the set of rotation matrices $D_{J,M,K}$:

$$|J,M,K\rangle = \sqrt{\frac{2J+1}{8\pi^2}}D^*_{J,M,K}(\theta,\varphi,\chi)$$

which obey,

$$J^2|J,M,K\rangle = \hbar^2 J(J+1)|J,M,K$$

$$J_a|J,M,K\rangle = \hbar K|J,M,K\rangle$$

$$J_z|J,M,K\rangle = \hbar M||J,M,K\rangle.$$

Symmetric Tops

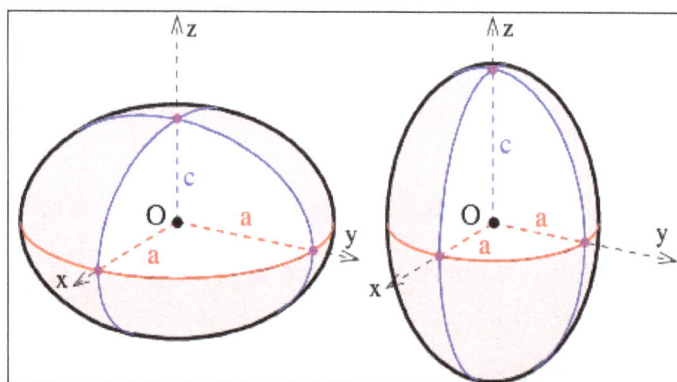

Figure: The assignment of semi-axes on a spheroid. It is oblate if c < a (left) and prolate if c > a (right).

Symmetrical tops are molecules with two rotational axes that have the same inertia and one unique rotational axis with a different inertia. Symmetrical tops can be divided into two categories based on the relationship between the inertia of the unique axis and the inertia of the two axes with equivalent inertia. If the unique rotational axis has a greater inertia than the degenerate axes the molecule is called an oblate symmetrical top. If the unique rotational axis has a lower inertia than the degenerate axes the molecule is called a prolate symmetrical top. For simplification think of these two categories as either frisbees for oblate tops or footballs for prolate tops.

Again, the rotational kinetic energy, which is the full rotational Hamiltonian, can be written in terms of the total rotational angular momentum operator J² and the component of angular momentum along the axis with the unique principal moment of inertia.

For prolate tops, Equation $H_{rot} = \dfrac{J_a^2}{2I_a} + \dfrac{J_b^2}{2I_b} + \dfrac{J_c^2}{2I_c}$ become,

$$H_{rot} = \dfrac{J^2}{2I} + J_a^2 \left(\dfrac{1}{2I_a} - \dfrac{1}{2I} \right)$$

For oblate tops, Equation $H_{rot} = \dfrac{J_a^2}{2I_a} + \dfrac{J_b^2}{2I_b} + \dfrac{J_c^2}{2I_c}$ becomes,

$$H_{rot} = \dfrac{J^2}{2I} + J_c^2 \left(\dfrac{1}{2I_c} - \dfrac{1}{2I} \right)$$

As a result, the eigenfunctions of H_{rot} are those of J² and J_a or J_c (and of J_z), and the corresponding energy levels.

The energies for prolate tops are,

$$E(J,K,M) = \dfrac{h^2 J(J+1)}{2I^2} + h^2 K^2 \left(\dfrac{1}{2I_a} - \dfrac{1}{2I} \right)$$

and the energies for oblate tops are,

$$E(J,K,M) = \dfrac{h^2 J(J+1)}{2I2} + h^2 K^2 \left(\dfrac{1}{2I_c} - \dfrac{1}{2I} \right)$$

again for K and M (i.e., J_a or J_c and J_z quantum numbers, respectively) ranging from $-J$ to J in unit steps. Since the energy now depends on K, these levels are only $2J+1$ degenerate due to the $2J+1$ different M values that arise for each J value. The eigenfunctions $\left| J, M, K \right\rangle$ are the same rotation matrix functions as arise for the spherical-top case.

Asymmetric Tops

The rotational eigenfunctions and energy levels of a molecule for which all three principal moments of inertia are distinct (a asymmetric top) can not easily be expressed in terms of the angular momentum eigenstates and the J, M, and K quantum numbers. However, given the three principal moments of inertia I_a, I_b, and I_c, a matrix representation of each of the three contributions to the general rotational Hamiltonian in Equation $H_{rot} = \dfrac{J_a^2}{2I_a} + \dfrac{J_b^2}{2I_b} + \dfrac{J_c^2}{2I_c}$ can be formed within a basis set of the $\{|J, M, K\rangle\}$ rotation matrix functions. This matrix will not be diagonal because the $|J, M, K\rangle$ functions are not eigenfunctions of the asymmetric top H_{rot}. However, the matrix can be formed in this basis and subsequently brought to diagonal form by finding its eigenvectors $\{C_n, J, M, K\}$ and its eigenvalues $\{E_n\}$. The vector coefficients express the asymmetric top eigenstates as,

$$\psi_n(\theta, \varphi, \chi) = \sum_{J, M, K} C_{n, J, M, K} |J, M, K\rangle$$

Because the total angular momentum J^2 still commutes with H_{rot}, each such eigenstate will contain only one J-value, and hence Ψ_n can also be labeled by a J quantum number:

$$\psi_{n, J}(\theta, \varphi, \chi) = \sum_{M, K} C_{n, J, M, K} |J, M, K\rangle$$

To form the only non-zero matrix elements of H_{rot} within the $|J, M, K\rangle$ basis, one can use the following properties of the rotation-matrix functions:

$$\langle j, \rangle = \langle j, \rangle = 1/2 \langle < j, \rangle = h2[J(J + 1) - k2],$$

$$\langle j, \rangle = h^2 K^2$$

$$\langle j, \rangle = -\langle j, \rangle = h^2 [J(J + 1) - K(K \pm 1)]1/2[J(J + 1) - (k \pm 1)(k \pm 2)]1/2 \langle j \rangle = 0.$$

Each of the elements of $J_c^2, J_a^2,$ and J_b^2 must, of course, be multiplied, respectively, by $1/2I_c, 1/2I_a$ and $1/2I_b$ and summed together to form the matrix representation of H_{rot}. The diagonalization of this matrix then provides the asymmetric top energies and wavefunctions.

References

- Rotational-spectroscopy, chemistry: sciencedirect.com, Retrieved 3 April, 2019

- Microwave-Rotational-Spectroscopy, Rotational-Spectroscopy, Spectroscopy: Physical-and-Theoretical-Chemistry: libretexts.org, Retrieved 11 January, 2019

- Rotational-Spectroscopy-of-Diatomic-Molecules, Rotational-Spectroscopy, Spectroscopy: Physical-and-Theoretical-Chemistry: libretexts.org, Retrieved 19 May, 2019

- A-Molecular-Spectroscopy, Physical-Chemistry: Physical-and-Theoretical-Chemistry: libretexts.org, Retrieved 9 July, 2019

3

Vibrational Spectroscopy

Vibrational spectroscopy is a method which is used to measure the vibrational energy of a compound. It is considered to be a non-destructive method of identification. The topics elaborated in this chapter will help in gaining a better perspective about the different aspects of vibrational spectroscopy such as isotope effects.

Vibrational spectroscopy is a non-destructive identification method that measures the vibrational energy in a compound.

Each chemical bond has a unique vibrational energy. Even a carbon-carbon bond will be different from one compound to another depending on what other compounds each carbon is bound to. Due to this unique vibrational energy, each compound will have a unique fingerprint, or the output identifying the peak strengths at specific vibrations. This fingerprint can be used to determine compound structures, identify and characterize compounds, and identify impurities. This is done by comparing the fingerprint with the fingerprints of known compounds.

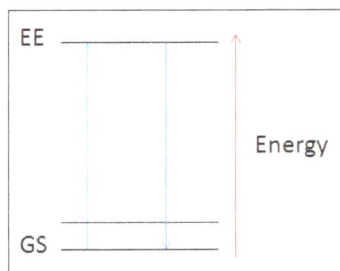
Figure: Electronic Excitation Diagram.

Vibrational spectroscopic methods use infrared or near infrared (the low energy end of the visible spectrum) to create vibrations (bond stretching or bending) in chemical species. Like visible spectroscopy, the radiation causing the vibration is absorbed and a corresponding peak is created on an Infrared or Raman spectrum. In figure below, the electronic ground state is again represented by GS and electronic excitation states are represented by EE. Vibrational excitation states are found at energy levels in between, notated by VE in the figure. The vibrational excitation states are lower in energy than electronic excitation states.

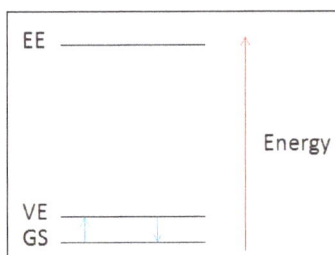
Figure: Vibrational Excitation Diagram. GS denotes the ground state of an electron, VE is a vibrational excitation state, and EE is an electronic excitation state.

The most widely used vibrational spectroscopy is Infrared (IR) spectroscopy. In IR spectroscopy, an infrared lamp produces electromagnetic radiation between the wavelengths of 700 nm to 1 mm. Converting wavelength to frequency $(v \; \alpha \; 1/\lambda)$, is convenient for this type of spectroscopy because vibrational excitations occur at characteristic frequencies in molecules. When a spectrum of a chemical is taken, the spectrometer scans through a range of wavelengths (frequencies). Electromagnetic interference is used to expose the sample with one wavelength (frequency) at a time. When a chemical sample is exposed to electromagnetic radiation at the right frequency, the atoms of the chemical vibrate at that frequency, resulting in the stretching of bond lengths and angles. The bond is analogous to a spring: when the bond is compressed or stretched an equal and opposing spring force is induced. This spring force is denoted by the Hooke's law equation:

$F = kx$

Where F is the spring force of the bond, k is the spring constant, and x is the distance between atomic nuclei.

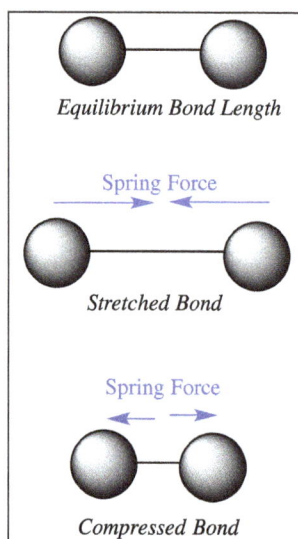

Figure: Vibrational Stretching Mechanisms.

The detector measures the reduction of the frequency of the electromagnetic radiation absorbed by the chemical sample, resulting in a peak on the spectrum. This peak occurs at this frequency due to the stretching and compressing of bonds in a molecule.

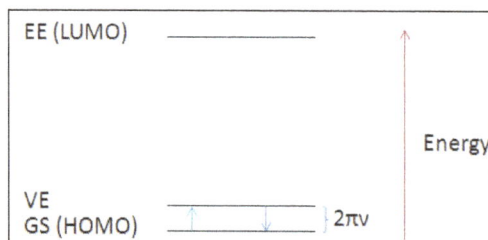

Figure: IR Vibrational Excitation Energy Diagram.

As shown in Figure, the energy states in the excitation diagram are labeled as GS (ground state), VE (vibrational excitation), and EE (electronic excitation) for simplicity. The ground state (GS) is the same as the highest occupied molecular orbital (HOMO) and the excited state (EE) is the same as the lowest unoccupied molecular orbital (LUMO). The energy of vibrational excitation states is found somewhere in between these orbitals.

To monitor vibrational excitations in a molecule, a Fourier Transform Infrared (FTIR) Spectrometer is most often used. The design of this instrument is shown in figure below. The excitation source is an *IR lamp* capable of emitting many wavelengths of radiation.

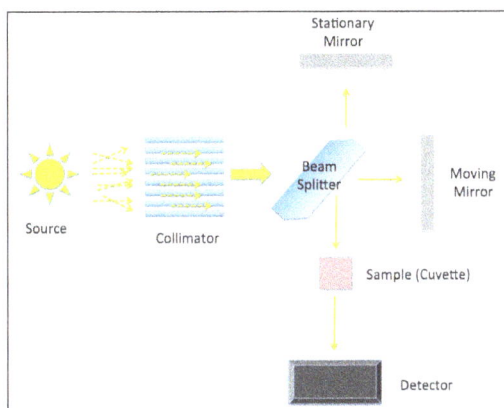

Figure: Instrumental Design of a FTIR spectrometer.

First, the radiation passes through a collimator to ensure the photons are all moving in the same direction, eliminating scattering. The radiation then enters a beam splitter and is separated into two beams. One beam is redirected at a 90° angle to a stationary mirror; the other is sent to an automated moving mirror. A time delay is created between the beams resulting in destructive interference between the two paths at certain wavelengths. As a result, only photons of a single wavelength reach the sample each time for excitation. To change the wavelength of the photons, the moving mirror is adjusted causing the interference to occur at the desired wavelength. By continually adjusting the moving mirror, the instrument can scan through the entire IR wavelength range to create a spectrum. The spectrum created is with respect to time. A Fourier transform (a mathematical operation that converts time to frequency) is performed to give spectral peak positions in wavenumbers (cm⁻¹), a frequency unit.

Now that the basics of IR spectroscopy have been covered, let's examine what happens at a more detailed level. As you know, a molecule is often composed of several different atoms bound to each other covalently. Each bond type in a molecule can be excited at a characteristic frequency. Table 1 contains some common vibrational mode frequencies for different bond types. When a bond in a molecule is vibrationally excited, a peak appears on the spectrum representing the absorption of energy at that frequency. Bond strength is directly related to frequency. In other words, more energy (a higher frequency) is required to vibrate a stronger bond. Because of this relationship, IR spectroscopy is typically used to show the presence of certain bonding types.

Table: Stretching Frequencies of Common Bonds.

Bond Type	Expected Frequency Range of peak (ω, cm^{-1})	Intensity (Peak Size)
O-H	3200-3700	strong
C-O	1050-1150	strong

C=O	1670-1820	strong
C-H	2850-3100	medium
C=C	1620-1680	variable
C≡C	2100-2260	variable
C-N	1250-1335	medium
C=N	2100-2270	medium
C≡N	2210-2260	medium
S-C	570-710	weak
S=C	1030-1200	strong

Absorption of IR radiation leads to the vibrational excitation of an electron. This excitation leads to the stretching and compressing of bonds. Vibrational excitations that change the bond dipole are IR active. For example, figure shows the bond dipoles (purple arrows) for a molecule of carbon dioxide in 3 different stretches/compressions. The top drawing shows the dipoles of CO_2 before the molecule is vibrationally excited. Once the molecule is vibrationally excited, the stretching/compressing of bonds can be symmetrical (as shown by the middle drawing) or unsymmetrical (as shown by the bottom drawing).

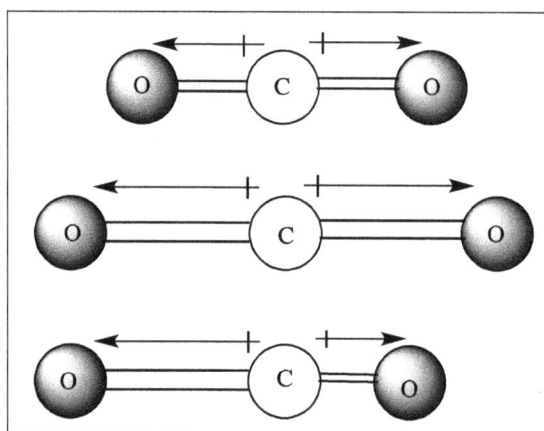

Figure: Bond Dipoles of CO_2.

If the stretch is symmetric, opposite and equal magnitudes of the dipole cancel each other out completely and the overall net dipole change is zero. However, if the stretch is asymmetric, the bond dipoles do not cancel each other out, resulting in a net dipole and an IR active mode. Therefore, polar molecules are always IR active and nonpolar molecules must vibrate in a nonsymmetrical way to be IR active.

Raman spectroscopy is another vibrational technique, currently gaining popularity because recent technological advances have made the instrumentation more accessible. Figure below shows a basic schematic of the Raman spectrometer you will use in the lab.

Raman spectroscopy uses a single wavelength laser source to excite the electrons in a sample. As shown in figure below, the laser source is aimed at a mirror that directs the beam to a polarizer and half-wave plate. Together these components act like a collimator like an IR spectrometer, ensuring all photons are moving in the same direction for sample excitation. The laser is then sent through a beam expander to ensure the spot size is a large enough to produce a strong enough response for measurement. The objective tightly focuses the expanded beam into a fiber optic cable that guides the laser to the sample. When the beam interacts with the sample, its electric field excites the molecules by inducing a change in their polarizability. (The 3 intramolecular mechanisms for this excitation are described below.) The radiation that has passed through the sample is then passed through a notch filter. The notch filter selectively blocks radiation at just the excitation beam frequency, leaving all other frequencies to pass through to the grating (monochromator) that spreads the radiation into separate frequencies for the detector.

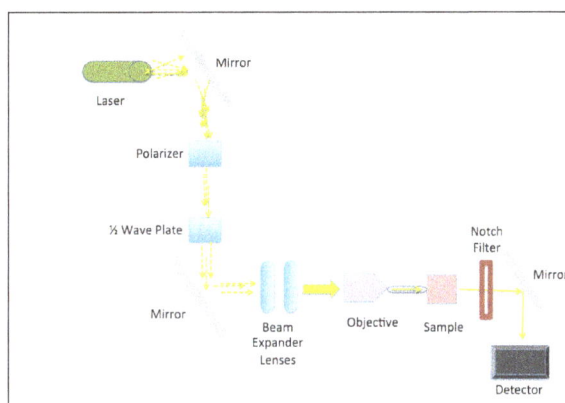

Figure: Instrumental Design of a Raman Spectrometer.

The excitation mechanisms for a change in polarizability are: Rayleigh scattering, Stokes scattering, and anti-Stokes scattering.

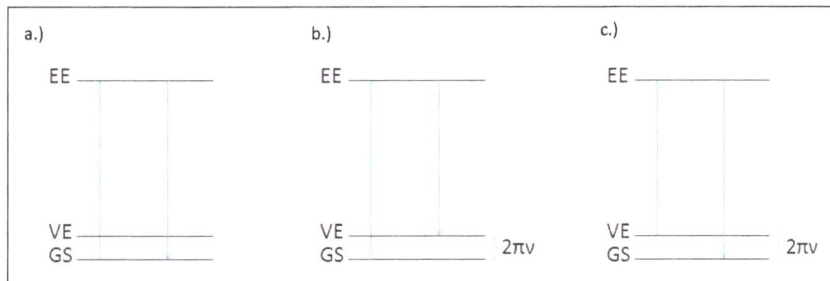

Figure: Excitation Diagrams for Rayleigh Scattering (a), Stokes Shift (b), and Anti-stokes Shift (c).

The mechanism in:

- Rayleigh scattering, is elastic because the excitation begins and ends in the same vibrational state. Elastic scattering of photons is far more likely than inelastic scattering of photons.

- Stokes and anti-Stokes scattering, is inelastic because the initial and final energy levels of the excited electron are different vibrational states. Stokes scattering is much more common that anti-Stokes scattering.

Stokes scattering, responsible for most peaks in a Raman spectrum, occurs by the following mechanism. When a laser beam interacts with a sample, an electronic excitation occurs from the ground state (GS) to the electronic excited state (EE). A change in polarizability of the molecule causes a small loss of energy. As a result, the electron relaxes down to the vibrational state (VE) (not the initial ground state (GS)). The frequency difference between VE and GS is equal to $2\pi v$ (v = frequency of the laser source). When this frequency difference is equal to the frequency at which a molecule can vibrate, a peak will be seen in spectrum at the corresponding wavenumber. This equality in frequency is known as resonance. The peak created is a Raman active peak and is reported in wavenumbers (cm-1) (just like the peaks in IR). As a general rule, an IR active vibrational mode is not Raman active and vice versa.

IR vs. Raman Spectroscopy

Raman spectroscopy is similar to IR spectroscopy in a few fundamental ways:

- Both techniques are measured in wavenumbers (cm⁻¹) because chemical bond vibration frequencies are being induced. So, typically the same chemical information is obtained from both techniques. In other words, the expected excitation frequencies for each bond type in a molecule occur at roughly the same wavenumber.

- Both techniques can be used to measure vibrational excitations for most matter in any physical state (solid, liquid, or gas).

- Raman spectroscopy differs from IR spectroscopy in a few fundamental ways.

- The excitation in Raman spectroscopy results in a transition between electronic states; in IR spectroscopy only a change in vibrational states occurs.

- The excitation source in Raman spectroscopy is a single wavelength (monochromatic) visible or near IR laser. The source in an IR spectrometer is capable of creating all frequencies of infrared radiation.

In Raman spectroscopy, electromagnetic radiation is not absorbed (as in IR spectroscopy), but scattered. The change in the intensity of radiation before and after the sample is detected. This difference is proportional to the frequency of the bond vibration. In IR spectroscopy a specific energy of electromagnetic radiation is absorbed by the sample, directly resulting in a peak on the spectrum.

Vibrational excitations that change bond polarizability are Raman active. Bond polarizability is the ease with which an electron cloud will move to one end of the bond. Nonpolar bonds are more polarizable than polar bonds. In a polar bond the more electronegative atom will hold the electrons closer to itself, limiting polarizability. This means nonpolar bonds are always Raman active and polar bonds decrease in Raman activity as the electronegativity difference in the atoms in the bond increases.

	Raman IR	Raman IR
1	Due to the scattering of light by the vibrating molecules.	Result of absorption of light by vibrating molecules.
2	The vibration is Raman active if it causes a change in polarizability.	The vibration is IR active if the dipole moment changes during the vibration.
3	The molecule need not possess a permanent dipole moment.	The vibration should have a dipole moment change due to that vibration.
4	Water can be used as a solvent.	Water generally cannot be used due to its intense absorption.

5	Sample preparation is generally not very elaborate; sample can be almost in any state. A solid sample can simply be placed on a glass slide without extra treatments Liquids and gases are simply placed in a cuvette for analysis. Liquids require small volume, gases require large volume. Why do you think this is???	Sample preparation is elaborate. For solids, the a sample must be ground together with KCl and mechanically pressed to form a pellet. Pellets can be very delicate and must be handled carefully to avoid sample destruction. Liquids and gases must be placed in special cells. These are very expensive to purchase and must be often customized for sample compatibility.
6	Indicates the covalent character in the molecule.	Indicates the ionic character in the molecule.

Isotope effects in Vibrational Spectroscopy

Isotopic substitution is a useful technique due to the fact that the normal modes of an isotopically substituted molecule are different than the normal modes of an unsubstituted molecule, leading to different corresponding vibrational frequencies for the substituted atoms. Vibrational spectroscopy is done in the infrared region of the electromagnetic spectrum, which ranges from around 10^{-6} to 10^{-3} meters. IR and Raman spectroscopy observe the vibrations of molecules, displaying the normal and local modes of the molecule in the spectra. Isotopes are atoms that share the same number of protons but differ in the number of neutrons contained in the nucleus, thus giving these atoms different mass numbers. The specific mass of each atom will affect the reduced mass of the overall molecule, therefore changing the vibrational frequencies of that molecule.

Diatomics

A diatomic molecule, as seen in figure below, contains two atoms, which can either be composed of the same or different elements. It is easier to focus on these types of molecules when analyzing and calculating vibrational frequencies because they are simpler systems than polyatomic molecules. Whether or not the diatomic consists of the same or different elements, a diatomic molecule will have only one vibrational frequency. This singular normal mode is because of the diatomic's linear symmetry, so the only vibration possible occurs along the bond connecting the two atoms.

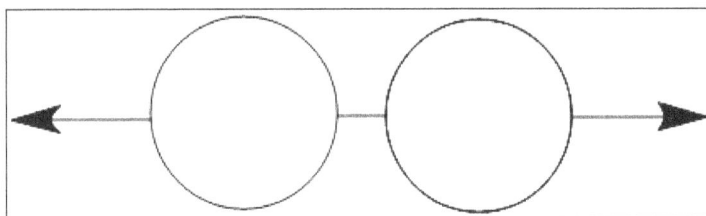

Figure: Diagram of a diatomic molecule with the only possible vibration it can undergo.

Normal Modes

Normal modes describe the possible movements/vibrations of each of the atoms in a system. There are many different types of vibrations that molecules can undergo, like stretching, bending, wagging, rocking, and twisting, and these types can either be out of plane, asymmetric, symmetric, or degenerate. Molecules have 3n possible movements due to their 3-dimensionality, where n is equal to the number of atoms in the molecule. Three movements are subtracted from the total because they are saved for the displacement of the center of mass, which keeps the distance and angles between the atoms constant. Another 3 movements are subtracted from the total because they are for the rotations about the 3 principle axes. This means that for nonlinear molecules, there are 3n − 6 normal modes possible. Linear molecules, however, will have 3n − 5 normal modes because it is not possible for internuclear axis rotation, meaning there is one less possible rotation for the molecule. This explain why diatomic molecules only have 1 vibrational frequency, because $3(2) - 5 = 1$.

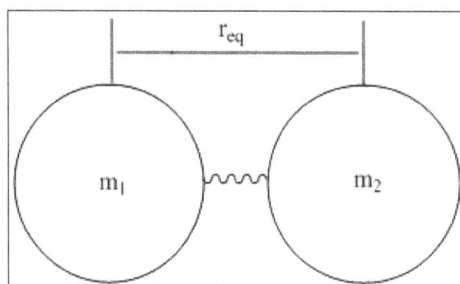

Figure: Representation of a diatomic molecule as masses attached by
a spring separated by an equilibrium distance of r_{eq}.

Molecular vibrations are often thought of as masses attached by a spring, and Hook's law can be applied,

$$F = -kx$$

where,

- F is the resulting force,

- x is the displacement of the mass from equilibrium $(x = r - r_{eq})$, and

- k is the force constant, defined as,

$$k = \left(\frac{\partial 2V(r)}{\partial r2} \right)_{req}$$

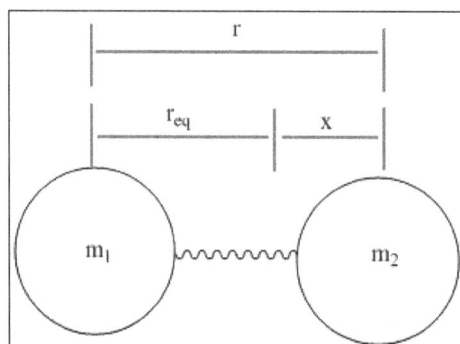

Figure: Depiction of the diatomic when the atoms undergo a vibration
and increase their separation by a distance x.

In which $V(r) = \frac{1}{2}k(r - r_{eq})$, which comes from incorporating Hook's law to the harmonic oscillator. The diatomic molecule is thought of as two masses (m_1 and m_2) on a spring, they will have a reduced mass, μ, so their vibrations can be mathematically analyzed.

$$\mu = \frac{m_1 m_2}{m_1 + m_2}$$

When an atom in a molecule is changed to an isotope, the mass number will be changed, so μμ will be affected, but k will not (mostly). This change in reduced mass will affect the vibrational modes of the molecule, which will affect the vibrational spectrum.

Vibrational energy levels, v_e, are affected by both k and μ, and is given by,

$$v_e = \frac{1}{2\pi}\sqrt{\frac{k}{\mu}}.$$

These vibrational energy levels correspond to the peaks which can be observed in IR and Raman spectra. IR spectra observe the asymmetric stretches of the molecule, while Raman spectra observe the symmetric stretches.

Effects on Experimental Results

When an atom is replaced by an isotope of larger mass, μ increases, leading to a smaller veve and a downshift (smaller wavenumber) in the spectrum of the molecule. Taking the diatomic molecule HCl, if the hydrogen is replaced by its isotope deuterium, μ is doubled and therefore veve will be decreased by $\sqrt{2}$. Deuterium substitution leads to an isotopic ratio of 1.35-1.41 for the frequencies corresponding to the hydrogen/deuterium vibrations. There will also be a decrease by $\sqrt{2}$ in the band width and integrated band width for the vibrational spectra of the substituted molecule. Isotopic substitution will affect the entire molecule (to a certain extent), so it is not only the vibrational modes for the substituted atom that will change, but rather the vibrational modes of all the atoms of the molecule. The change in frequency for the atoms not directly invovled in the substitution will not display as large a change, but a downshift can still occur.

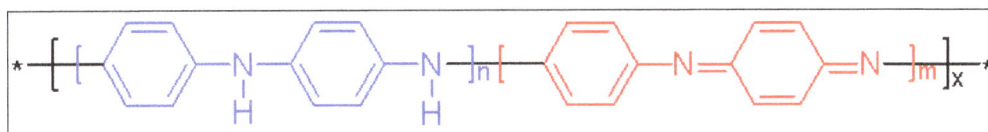

Figure: Structure of polyaniline.

When polyaniline is fully deuterated, the vibrational peaks will downshift slightly. The following data was summarized from Quillard *et al.*

Type of vibration	Nondeuterated (frequency, cm⁻¹)	Deuterated (frequency, cm⁻¹)
C-C stretch	1626	1599

C-C stretch	1581	1560
C-H bend – benzenoid ring	1192	876
C-H bend – quinoid ring	1166	856
N-H bend	1515	1085

Changing hydrogen to deuterium leads to the largest effect in a vibrational spectrum since the mass is doubled. Other isotopic substitutions will also lead to a shift in the vibrational energy level, but because the mass change is not as significant, μ will not change by much, leading to a smaller change in veve. This smaller change in vibrational frequency is seen in the sulfur substitution of sulfur hexafluoride, from ^{32}S to ^{34}S. The frequencies as reported by Kolomiitsova et $al.$ are shown below.

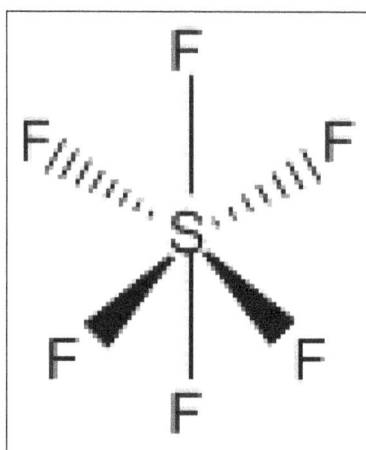

Figure: Structure of sulfur hexafluoride.

Vibration Assignment	$^{32}SF_6$ (frequency, cm^{-1})	$^{34}SF_6$ (frequency, cm^{-1})
v3	939.3	922.2
v4	613.0	610.3

These two examples show the consistency of downshifted vibrational frequencies for atoms substituted with an isotope of higher mass.

Applications

Substituting atoms with isotopes has been shown to be very useful in determining normal mode vibrations of organic molecules. When analyzing the spectrum of a molecule, isotopic substitution can help determine the vibrational modes specific atoms contribute to. Those normal modes can be assigned to the peaks observed in the spectrum of the molecule. There are specific CH_3 rocks and torsions, as well as CH bends that can be identified in the spectrum upon deuterium substitution. Other torsion bands from hydroxyl and amine groups can also be assigned when hydrogen is replaced with deuterium. Experimental data has also shown that using deuterium substitution can help with symmetry assignments and the identification of metal hydrides.

Researchers have also attempted to contribute peak shape changes and splits in peaks of vibrational spectra to naturally occurring isotopes in molecules. However, that the shape of a peak is not related to the size of the atom, so substitution to an atom of larger mass will not affect the peak shape in the molecule's spectrum. As previously stated, isotopic substitution of atoms of higher mass will not have a significant enough effect on the shifts in frequencies for the corresponding vibrations, so analyzing the frequency shifts of smaller mass isotopes, like deuterium and ^{13}C is necessary.

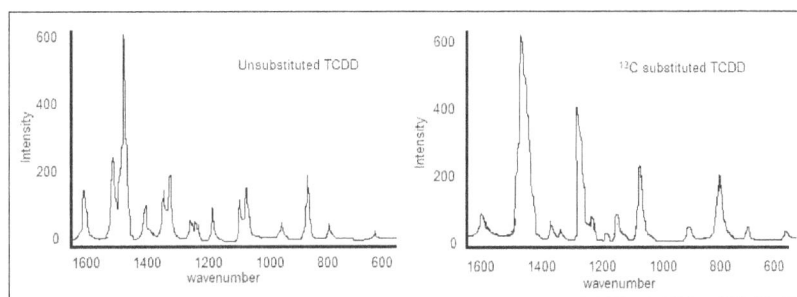

Figure: Spectra of unsubstituted and substituted TCDD depicting the isotopic 13C effects.

As depicted in the rough representation of the vibrational spectra of the molecule tetrachlorinated dibenzodioxin (TCDD), the ^{13}C substituted TCDD spectrum is slightly downshifted compared to the unsubstituted TCDD spectrum. Although the shifts and split peaks do occur in the spectra of isotopically substituted molecules, not all observed peaks can be attributed to the isotope. This is because the intensities of the peaks shown are not large enough to relate to the natural abundance of the ^{13}C isotope, and not all peaks can be accounted for by the substitution.

Vibrational Spectroscopy of Diatomic and Polyatomic Molecules

A diatomic molecule has 1 vibrational degree of freedom that relates to the stretching of the bond. A polyatomic molecule with N atoms has 3N6(nonlinear) or 3N-5(linear) vibrational degrees of freedom. This vibrational motion of the molecule is probed by vibrational spectroscopy.

The vibrational energy levels of the molecule can be easily expressed if we treat the vibrations using the harmonic oscillator model. According to this model, the vibrational energy levels are given by,

$$E_v = (v + \frac{1}{2}) h v$$

where nu is the vibrational frequency and is given by p $\sqrt{k / \mu}$ where k is the spring constant of the bond and v is the reduced mass of the molecule. Thus, if the spring constant of the bond is known, then the energy levels of the molecule can be calculated.

The frequency of light corresponding to vibrational energies corresponds to the IR (infra-red) light. Typically the wavenumber is used with units of cm^{-1}.

The gross selection rule for a molecule to be IR active and show vibrational transitions is that it should have a dipole moment that changes as the molecule vibrates. For a diatomic molecule, there is only one vibrational mode. If the molecule is homonuclear, then it cannot be IR active whereas if it is heteronuclear, it can be IR active. On the other hand, polyatomic molecules have several vibrational modes which may or may not be IR active. For example, water has 3 vibrational modes corresponding to symmetric stretch, asymmetric stretch and one bends. All three modes are IR active. For CO_2 (linear molecule) there are 4 vibrational modes corresponding to symmetric stretch, antisymmetric stretch and two bends. The symmetric stretch does not change the dipole moment so it is not IR active.

The specific selection rule for a vibrational transition derived as per the Harmonic oscillator model is that $\Delta v = \pm 1$. Since the spacing between any two energy levels in a harmonic oscillator is equal to hvo, where vo is the oscillator frequency, the frequency of light should equal the frequency of the oscillator.

$$hv_l = hv_o \quad v_l = v_o = \frac{1}{2\pi}\sqrt{\frac{k}{\mu}}$$

Thus, if we know the frequency of light absorbed, we can calculate the oscillator frequency and find the spring constant of the bond.

IR spectroscopy is one of the most widely used spectroscopic techniques for organic molecules.

4

Electron Spectroscopy

Electron spectroscopy deals with the measurement of the kinetic energies which are emitted by electrons when they are bombarded with UV or X-ray radiation. It is used to determine the energy with which electrons are bound in chemical species. This chapter has been carefully written to provide an easy understanding of the varied techniques used within electron spectroscopy such as Auger Electron spectroscopy and electron energy loss spectroscopy.

Electron spectroscopy is a method of determining the energy with which electrons are bound in chemical species by measuring the kinetic energies of the electrons emitted upon bombardment of the species with X-ray or ultraviolet radiation. Details of the structure may be inferred from the results because differences in the arrangements of the atoms affect the amount of energy required to eject electrons. The most common type is photoelectron spectroscopy, in particular X-ray photoelectron spectroscopy (XPS) but also the UV photoelectron spectroscopy (UPS) is widely used. In most XPS spectra also Auger precesses are seen. Even the electron enrgy loss spectroscopy (EELS) can give valuable information. The instruments for photoelectron spectroscopy are often equipped to show also other kinds of spectra such as secondary ion mass spectra (SIMS).

Photoelectron Spectroscopy

Photoelectron spectroscopy (PES) is a technique used for determining the ionization potentials of molecules. Underneath the banner of PES are two separate techniques for quantitative and qualitative measurements. They are ultraviolet photoeclectron spectroscopy (UPS) and X-ray photoelectron spectroscopy (XPS). XPS is also known under its former name of electron spectroscopy for chemical analysis (ESCA). UPS focuses on inoization of valence electrons while XPS is able to go a step further and ionize core electrons and pry them away.

Photoelectron Instrumentation

The main goal in either UPS or XPS is to gain information about the composition, electronic state, chemical state, binding energy, and more of the surface region of solids. The key point in PES is that a lot of qualitative and quantitative information can be learned about the surface region of solids. The goal is to understand how to go about constructing or diagramming a PES instrument, how to choose an appropriate analyzer for a given system, and when to use either XPS or UPS to study a system.

There are a few basics common to both techniques that must always be present in the instrumental setup.

- A radiation source: The radiation sources used in PES are fixed-energy radiation sources. XPS sources from x-rays while UPS sources from a gas discharge lamp.

- An analyzer: PES analyzers are various types of electron energy analyzers.

- A high vacuum environment: PES is rather picky when it comes to keeping the surface of the sample clean and keeping the rest of the environment free of interferences from things like gas molecules. The high vacuum is almost always an ultra high vacuum (UHV) environment.

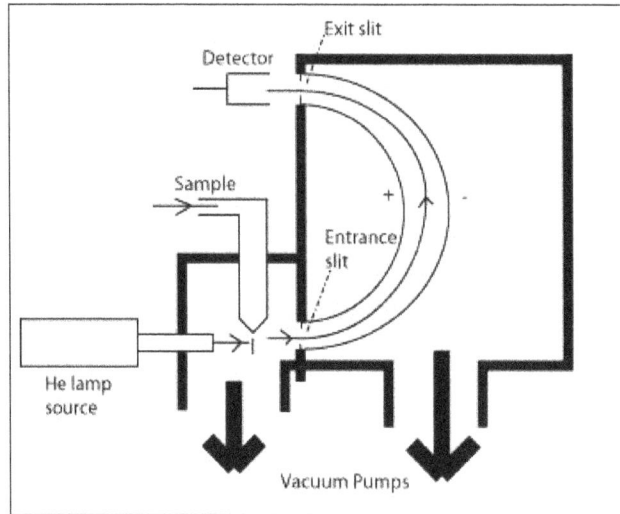

Above diagram of a basic, typical PES instrument used in XPS, where the radiation source is an X-ray source. When the sample is irradiated, the released photoelectrons pass through the lens system which slows them down before they enter the energy analyzer. The analyzer shown is a spherical deflection analyzer which the photoelectrons pass through before they are collected at the collector slit.

Radiation Sources

While many components of instruments used in PES are common to both UPS and XPS, the radiation sources are one area of distinct differentiation. The radiation source for UPS is a gas discharge lamp, with the typical one being an He discharge lamp operating at 58.4 nm which corresponds to 21.2 eV of kinetic energy. XPS has a choice between a monocrhomatic beam of a few microns or an unfocused non-monochromatic beam of a couple centimeters. These beams originate from X-Ray sources of either Mg or Al K-? sources giving off 1486 eV and 1258 eV of kinetic energy respectively. For a more versitile light source, synchrotron radiation sources are also used. Synchrotron radiation is especially useful in studying valence levels as it provides continuous, polarized radiation with high energies of > 350 eV.

The main thing to consider when choosing a radiation source is the kinetic energy involved. The source is what sets the kinetic energy of the photoelectrons, so there needs to not only be enough energy present to cause the ionizations, but there must also be an analyzer capable of measuring the kinetic energy of the released photoelectrons.

In XPS experiments, electron guns can also be used in conjunction with x-rays to eject photoelectrons. There are a couple of advantages and disadvantages to doing this, however. With an electron gun, the electron beam is easily focused and the excitation of photoelectrons can be constantly varied. Unfortunately, the background radiation is increased significantly due to the scattering of falling electrons. Also, a good portion of substances that are of any experimental interest are actually decomposed by heavy electron bombardment such as that coming from an electron gun.

Analyzers

There are two main classes of analyzers well-suited for PES - kinetic energy analyzers and deflection or electrostatic analyzers. Kinetic energy analyzers have a resolving power of $E/\delta E$, which means the higher the kinetic energy of the photoelectrons, the lower the resolution of the spectra. Deflection analyzers are able to separate out photoelectrons through an electric field by forcing electrons to follow different paths according to their velocities, giving a resolving power $E/\delta E$, that is greater than 1,000.

Since the resolving power of both types of analyzer is $E/\delta E$, the resolution is directly dependent on the kinetic energy of the photoelectrons. The intensity of the spectra produced is also dependent on the kinetic energy. The faster the electrons are moving, the lower the resolution and intensity is. In order to actually get well resolved, useful data other components must be introduced into the instrument.

Adding a system of optics (lenses) to a PES instrument helps with this problem immensely. Electron optics are capable of decelerating the photoelectrons through retardation of the electric field. The energy the photoelectrons decelerate to is known as the "pass energy." This has the benefit of significantly raising the resolution, however this does, unfortunately, lower the sensitivity. Optics are also capable of accelerating the electrons as well. The design of any lens system greatly effects the photoelectron counts. These lenses are also capable of focusing on a small area of a particular sample.

Specific Analyzers

Within the broad picture of two main analyzer classes, there are a variety of specific analyzers in existence that are used in PES. The list below goes over several well-used analyzers, though this list is, by no means, exhaustive. The most common type of analyzer is a hemispherical analyzer, which will be explained in more depth under the spherical deflection analyzer topic.

Plane Mirror Analyzer (PMA)

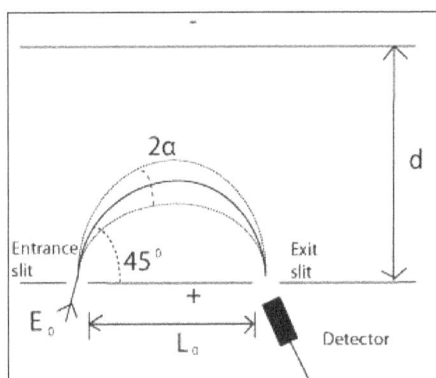

Schematic of a PMA where the angle between the bottom plate and the electrons entering is 45° and the angle between the bottom plate and the electrons exiting is also 45 degrees.

PMAs, the simplest type of electric analyzer are also known as parallel-plate mirror analyzers. These analyzers are condensers made from two parallel plates with a distance, d, across them. Parabolic trajectories of electrons are obtained due to the constant potential difference, V, between the two plates.

In order for transmission to occur, the potential must be: $V=E_{od}/eL_o$. E_o=kinetic energy of electron in eV and e=charge of the electron.

To obtain better focus, the electron entrance and exit angle is capable of being shifted to 30 degrees, but this is not necessarily a good idea as it sacrifices transmission instead.

Cylindrical Mirror Analyzer (CMA)

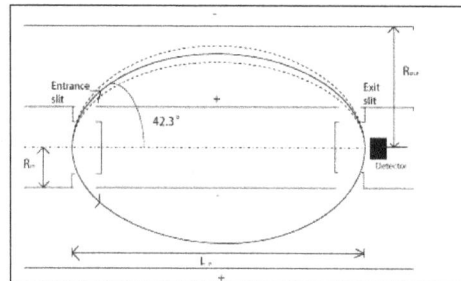

Schematic of a CMA where the angle between the center of the cylinders and the electrons is 42.3 degrees. R_{in} is the radius of the inner cylindar and Rout is the radius of the outer cylinder. The electron path should be more parabolic than the overly elliptical shape shown here.

CMAs are advantageous over PMAs. They employ to geometry to overcome the low transmission with a PMA. A CMA consists of two cylinders having a potential difference, V, between them. The entrance and exit slits are all contained on the inner cylinder.

Here: $V = 1.3E_0 \ln(R_{out}/R_{in})$ where $L_0 = 6.1(R_{in})$ and E_0 is in volts.

They are good for applications that require a high sensitivity with only a moderate resolution.

Cylindrical Deflection Analyzer (CDA)

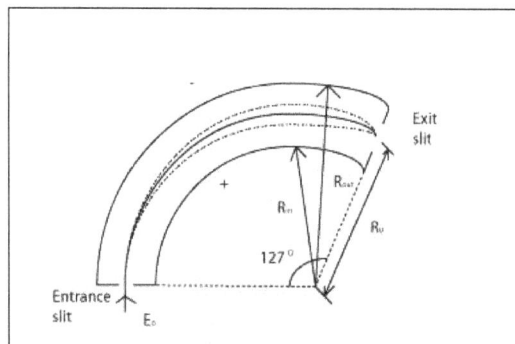

Figure: Schematic of a CDA where the angle the cylinders span is 127 degrees.

CDAs consist of two cylinders spanning a 127 degree angle. It is this reason that CDAs are sometimes called "127 degree analyzers."

- The potential difference in a CDA is: $2V = E_0(R_{in}/R_{out})$ where E_o is the energy of incoming photoelectrons, in eV, that are focused.

These analyzers have high resolution, however their transmission is low.

Spherical Deflection Analyzer (SDA)

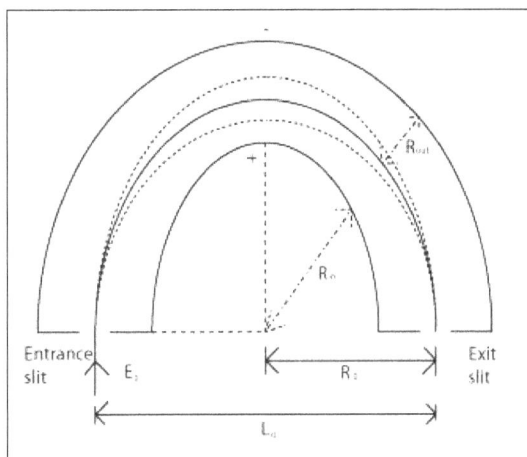

Figure: Schematic of an SDA. Only photoelectrons of the correct energy are able to pass through the detector with the right arc and exit instead of colliding with the side walls of the hemispheres and becoming lost.

Since SDAs are the most common, prevalent type of PES analyzer, than any of the previous analyzers as a thorough understanding of how they apply to PES is, theoretically, of greater importance.

SDAs are similar to CDAs, but they consist of two concentric hemispheres instead. In an SDA, the transmission of photoelectrons with initial energy, E_o, occurs along a paty where $R_0 = \left(R_{in} / R_{out} \right) / 2.$

Here, the potential is different for both the inner and the outer hemisphere: $V_{in} = E_o [3 - 2(R_o/R_{in})]$ and $V_{out} = E_o [3 - 2 (R_o/R_{out})].$

The resolving power of these analyzers is proportional to the radius of the inner and outer hemispheres. These analyzers are aslo capable of running in two separate modes when coupled with an optical system - fixed analyzer transmission mode (FAT) and fixed retardation energy mode (FRR). In FAT mode, the lens either retards or accelerates the electrons so that all photoelectrons enter the analyzer with the same kinetic energy. For this to occur, the analyzer is alos arranged so that only photoelectrons of a specific, fixed kinetic energy will pass through and reach the detector. In this case, the lens is scanned for different energies. In FRR mode, the lens only retards the photoelectrons, and it does so in a uniform manner causing all photoelectrons to be reduced in energy to a fixed value such as 15 eV, 30 eV, or whatever energy is desired. The hemispheres of the analyzer here have a potential difference between them that is varied so that photoelectrons of different kinetic energies can reach the detector. The more common of these two modes is FAT because it provides a greater signal intensity at low elecron kinetic energy and is also makes quantification of the spetra simpler.

These analyzers are a particularly good class of deflection analyzer. The slits in an SDA define the acceptable range of entrance and exit trajectories a photoelectron may have when entering or leaving the analyzer. The photoelectrons that do make it through the entrance slits will then only exit if they follow a specific, curved path down the middle of the two hemispheres. The path they follow has the "correct energy" for exit to occur, and is determined by the selection of V_{in} and V_{out}. Photoelectrons that are of higher or lower kinetic energy than what is defined by the hemispheres will be lost through collisions with the walls.

Detection and Spectra

Detection relies on the ability of the instrument to measure energy and photoelectron output. One type of energy measured is the binding energy, which is calculated through the following equation:

$$K_e = h\nu - BE - \phi$$

where:

- K_e = Kinetic energy, this is measured.

- $h\nu$ = Photon energy from the radiation source, this is controlled by the source.

- ϕ = Work function of the spectrometer, this is found through calibration.

- BE = Binding energy, this is the unknown of interest and can be calculated from the other three variables.

Another part of PES detection is in the use of electron multipliers. These devices act as electron amplifiers because they are coated with a material that produce secondary photoelectrons when they are struck by an electron. Typically, they are able to produce two to three photoelectrons per every electron they are hit with. Since the signals in PES are low, the huge amplification, up to 107 and higher when run in series so the secondary electrons from one multiplier strike the next, they greately improve the signal strength from these instruments.

One type of spectra in these experiments is recorded by varying the potential difference between the plates or hemispheres of the analyzer. The output is known as an electron kinetic energy spectrum and is obtained by measuring the photoelectron current at the detector as a function of the voltage applied to the hemispheres or plates. The voltage is then used in the calculation of kinetic energy.

Further detail on the spectra produced in PES experiments and the analysis of said spectra is planned for a future module on the interpreation of photoelectron spectroscopy.

Limitations

The main limitation in a PES instrument is the resolution. The problems in resolution come from four main areas: the dimensions of the analyzer, the widths of the entrance and exit slits, other charges such as outside electronic fields or outside magnetic fields, and local charges inside the instrument itself arising from things such as contamination in the analyzer. Steps can be taken to improve the resolution, but some methods then sacrifice other factors such as the sensitivity. Obtaining high spatial resolution and high energy resolution always comes at the expense of the signal intensity.

One resolution improving technique that, then, messes with the sensitivity is changing the width of the entrance and/or exit slits. For example, in an SDA these slits are what define the range of trajectories photoelectrons may have when entering or exiting the analyzer. Decreasing the widths will certainly cause the resolution to go up, but the smaller slit size will decrease the number of photoelectrons allowed in and out of the analyzer, therefore lowering the sensitivity.

A third method of improving resolution is specific to XPS and is the addition of an x-ray monochromator to the system. These monochromators eliminate satellite radiation from x-rays and are capable of narrowing the x-ray line width from $\sim 1\,eV$ to $\sim 0.2\,eV$. The use of monochromatic x-rays also serves to simplify the spectrum.

X-ray Photoelectron Spectroscopy

X-ray photoelectron spectroscopy, which is also known as Electron Spectroscopy for Chemical Analysis (ESCA), provides quantitative compositional information from the top atomic layers of a sample surface for the elements lithium to uranium. The center can also obtain information regarding the chemical states of any elements present.

XPS is used to analyze the change in surface chemistry of a material after certain chemical or physical treatments such as:

- Leaching;

- Reduction;

- Fracture;

- Cutting or scraping in air or UHV to expose the bulk chemistry;

- Ion beam etching to clean off some of the surface contamination;

- Exposure to heat to study the changes due to heating;

- Exposure to reactive gases or solutions;

- Exposure to ion beam implant and, exposure to UV light.

How Measurements are Obtained?

A sample is irradiated with a beam of monochromatic soft X-rays. Photoelectron emission results from the atoms in the specimen. The kinetic energies of these electrons relates to the atom and orbital from which they originated. The distribution of kinetic energies from a sample is then measured directly by the electron spectrometer.

Atomic orbitals from atoms of the same element in different chemical environments are found to possess slightly different (but measurable) binding energies. These "chemical shifts" arise because of the variations in electrostatic screening experienced by core electrons as the valence and conduction electrons are drawn towards or away from the specific atom. Differences in oxidation state, molecular environment and co-ordination number all provide different chemical shifts.

Photoelectron binding energy shifts are, therefore, the principal source of chemical information. It should be noted that these shifts can be very small and can only be detected using a high performance instrument with suitable software such as our Kratos Axis Ultra DLD instrument equipped with a monochromated Al Ka x-ray source and hemispherical analyzer.

XPS can be performed using a commercially built XPS system, a privately built XPS system, or a synchrotron-based light source combined with a custom-designed electron energy analyzer. Commercial XPS instruments in the year 2005 used either a focused 20- to 500-micrometer-diameter beam of monochromatic Al K_α X-rays, or a broad 10- to 30-mm-diameter beam of non-monochromatic (polychromatic) Al K_α X-rays or Mg K_α X-rays. A few specially designed XPS instruments can analyze volatile liquids or gases, or materials at pressures of roughly 1 torr (1.00 torr = 1.33 millibar), but there are relatively few of these types of XPS systems. The ability to heat or cool the sample during or prior to analysis is relatively common.

Because the energy of an X-ray with particular wavelength is known (for Al K_α X-rays, E_{photon} = 1486.7 eV), and because the emitted electrons' kinetic energies are measured, the electron binding energy of each of the emitted electrons can be determined by using an equation that is based on the work of Ernest Rutherford:

$$E_{binding} = E_{photon} - \left(E_{kinetic} + \phi \right)$$

where $E_{binding}$ is the binding energy (BE) of the electron, E_{photon} is the energy of the X-ray photons being used, $E_{kinetic}$ is the kinetic energy of the electron as measured by the instrument and ϕ is the work function dependent on both the spectrometer and the material. This equation is essentially a conservation of energy equation. The work function term conservation of energy equation. The work function is an adjustable instrumental correction factor that accounts for the few eV of kinetic energy given up by the photoelectron as it becomes absorbed by the instrument's detector. It is a constant that rarely needs to be adjusted in practice.

Basic Physics

A typical XPS spectrum is a plot of the number of electrons detected (sometimes per unit time) (Y-axis, ordinate) versus the binding energy of the electrons detected (X-axis, abscissa). Each element produces a characteristic set of XPS peaks at characteristic binding energy values that directly identify each element that exists in or on the surface of the material being analyzed. These characteristic spectral peaks correspond to the electron configuration of the electrons within the atoms, e.g., 1s, 2s, 2p, 3s, etc. The number of detected electrons in each of the characteristic peaks is directly related to the amount of element within the XPS sampling volume. To generate atomic percentage values, each raw XPS signal must be corrected by dividing its signal intensity (number of electrons detected) by a "relative sensitivity factor" (RSF), and normalized over all of the elements detected. Since hydrogen is not detected, these atomic percentages exclude hydrogen.

To count the number of electrons during the acquisition of a spectrum with a minimum of error, XPS detectors must be operated under ultra-high vacuum (UHV) conditions because electron counting detectors in XPS instruments are typically one meter away from the material irradiated with X-rays. This long path length for detection requires such low pressures.

Surface Sensitivity

XPS detects only those electrons that have actually escaped from the sample into the vacuum of the instrument, and reach the detector. In order to escape from the sample into vacuum, a photo-electron must travel through the sample. Photo-emitted electrons can undergo inelastic collisions, recombination, excitation of the sample, recapture or trapping in various excited states within the material, all of which can reduce the number of escaping photoelectrons. These effects appear as an exponential attenuation function as the depth increases, making the signals detected from analytes at the surface much stronger than the signals detected from analytes deeper below the sample surface. Thus, the signal measured by XPS is an exponentially surface-weighted signal, and this fact can be used to estimate analyte depths in layered materials.

Components of a Commercial System

An inside view of an old-type, non-monochromatic XPS system.

The main components of a commercially made XPS system include a source of X-rays, an ultra-high vacuum (UHV) stainless steel chamber with UHV pumps, an electron collection lens, an electron energy analyzer, Mu-metal magnetic field shielding, an electron detector system, a moderate vacuum sample introduction chamber, sample mounts, a sample stage, and a set of stage manipulators.

Monochromatic aluminum K_α X-rays are normally produced by diffracting and focusing a beam of non-monochromatic X-rays off of a thin disc of natural, crystalline quartz with a <1010> orientation. The resulting wavelength is 8.3386 angstroms (0.83386 nm) which corresponds to a photon energy of 1486.7 eV. Aluminum K_α X-rays have an intrinsic full width at half maximum (FWHM) of 0.43 eV, centered on 1486.7 eV ($E/\Delta E = 3457$). For a well–optimized monochromator, the energy width of the monochromated aluminum K_α X-rays is 0.16 eV, but energy broadening in common electron energy analyzers (spectrometers) produces an ultimate energy resolution on the order of FWHM=0.25 eV which, in effect, is the ultimate energy resolution of most commercial systems. When working under practical, everyday conditions, high energy-resolution settings will produce peak widths (FWHM) between 0.4–0.6 eV for various pure elements and some compounds. For example, in a spectrum obtained in 1 minute at a pass energy of 20 eV using monochromated aluminum K_α X-rays, the Ag $3d_{5/2}$ peak for a clean silver film or foil will typically have a FWHM of 0.45 eV.

Non-monochromatic magnesium X-rays have a wavelength of 9.89 angstroms (0.989 nm) which corresponds to a photon energy of 1253 eV. The energy width of the non-monochromated X-ray is roughly 0.70 eV, which, in effect is the ultimate energy resolution of a system using non-mono-chromatic X-rays. Non-monochromatic X-ray sources do not use any crystals to diffract the X-rays which allows all primary X-rays lines and the full range of high-energy Bremsstrahlung X-rays (1–12 keV) to reach the surface. The ultimate energy resolution (FWHM) when using a non-mono-chromatic Mg K_α source is 0.9–1.0 eV, which includes some contribution from spectrometer-induced broadening.

Uses and Capabilities

XPS is routinely used to determine:

- What elements and the quantity of those elements that are present within the top 1-12 nm of the sample surface.

- What contamination, if any, exists on the surface or in the bulk of the sample.

- Empirical formula of a material that is free of excessive surface contamination.

- The chemical state identification of one or more of the elements in the sample and also give information on local bonding of atoms.

- The binding energy of one or more electronic states.

- The thickness of one or more thin layers (1–8 nm) of different materials within the top 12 nm of the surface.

- The density of electronic states.

Capabilities of Advanced Systems

- Measure uniformity of elemental composition across the top of the surface (or line profiling or mapping).

- Measure uniformity of elemental composition as a function of depth by ion beam etching (or depth profiling).

- Measure uniformity of elemental composition as a function of depth by tilting the sample (or angle-resolved XPS).

Chemical States and Chemical Shift

The ability to produce chemical state information (as distinguished from merely elemental information) from the topmost few nm of any surface makes XPS a unique and valuable tool for understanding the chemistry of a surface. In this context, "chemical state" refers to the local bonding environment of a species in question. The local bonding environment of a species in question is affected by its formal oxidation state, the identity of its nearest-neighbor atom, its bonding hybridization to that nearest-neighbor atom, and in some cases even the bonding hybridization between the atom in question and the next-nearest-neighbor atom. Thus, while the nominal binding energy of the C_{1s} electron is 284.6 eV (some also use 285.0 eV as the nominal value for the binding energy of carbon), subtle but reproducible shifts in the actual binding energy, the so-called *chemical shift* (analogous to NMR spectroscopy), provide the chemical state information referred to here.

Chemical-state analysis is widely used for the element carbon. Chemical-state analysis of the surface of carbon-containing polymers readily reveals the presence or absence of the chemical states of carbon shown in bold, in approximate order of increasing binding energy, as: carbide (C^{2-}), silane (-Si-CH$_3$), methylene/methyl/hydrocarbon (-CH$_2$-CH$_2$-, CH$_3$-CH$_2$-, and -CH=CH-), amine (-CH$_2$-NH$_2$), alcohol (-C-OH), ketone (-C=O), organic ester (-COOR), carbonate (CO_3^{2-}), mono-fluoro-hydrocarbon (-CFH-CH$_2$-), difluoro-hydrocarbon (-CF$_2$-CH$_2$-), and trifluorocarbon (CH$_2$-CF$_3$), to name but a few examples.

Chemical state analysis of the surface of a silicon wafer readily reveals chemical shifts due to the presence or absence of the chemical states of silicon in its different formal oxidation states, such as: n-doped silicon and p-doped silicon, silicon suboxide (Si$_2$O), silicon monoxide (SiO), Si$_2$O$_3$, and silicon dioxide (SiO$_2$). An example of this is seen in the figure: High-resolution spectrum of an oxidized silicon wafer in the energy range of the Si 2*p* signal.

Routine Limits

Quantitative Accuracy and Precision

- XPS is widely used to generate an empirical formula because it readily yields excellent quantitative accuracy from homogeneous solid-state materials.

- Quantification can be divided into two categories: absolute quantification and relative quantification. The former generally requires the use of certified (or independently verified) standard samples, is generally more challenging, and is generally less common.

- Relative quantification is more common and involves comparisons between several samples in a set for which one or more analytes are varied while all other components (the sample matrix) are held constant.

- Quantitative accuracy depends on several parameters such as: signal-to-noise ratio, peak intensity, accuracy of relative sensitivity factors, correction for electron transmission function, surface volume homogeneity, correction for energy dependence of electron mean free path, and degree of sample degradation due to analysis.

- Under optimum conditions, the quantitative accuracy of the atomic percent (at%) values calculated from the Major XPS Peaks is 90-95% for each major peak. If a high level quality control protocol is used, the accuracy can be further improved.

- Under routine work conditions, where the surface is a mixture of contamination and expected material, the accuracy ranges from 80-90% of the value reported in atomic percent values.

- The quantitative accuracy for the weaker XPS signals, that have peak intensities 10-20% of the strongest signal, are 60-80% of the true value, and depend upon the amount of effort used to improve the signal-to-noise ratio (for example by signal averaging).

- Quantitative precision (the ability to repeat a measurement and obtain the same result) is an essential consideration for proper reporting of quantitative results. Standard statistical tests, such as the Student's t test for comparison of means, should be used to determine confidence levels in the average value from a set of replicate measurements, and when comparing the average values of two or more different sets of results. In general, a p value (an output of the Student's t test) of 0.05 or less indicates a level of confidence (95%) that is accepted in the field as significant.

Analysis Time

Typically ranging 1–20 minutes for a broad survey scan that measures the amount of all detectable elements, typically 1–15 minutes for high resolution scan that reveal chemical state differences (for a high signal/noise ratio for count area result often requires multiple sweeps of the region of interest), 1–4 hours for a depth profile that measures 4–5 elements as a function of etched depth (this process time can vary the most as many factors will play a role).

Detection Limits

Detection limits may vary greatly with the cross section of the photoelectron line of interest and the background signal level which is a function of the matrix material. In general photoelectron cross sections increase with atomic number, while the background is a function of the composition of the matrix material and the binding energy. Background signals generally increase with atomic number of the matrix material and decrease with increasing kinetic energy. For example in the case of gold on silicon where the high cross section. Au4f peak is at a higher kinetic energy than the major silicon peaks, it sits on a very low background and detection limits of 1ppm or better may be achieved with reasonable acquisition times. Conversely for silicon on gold, where the modest cross section Si2p line sits on the large background below the Au4f lines, detection limits would be much worse for the same acquisition time. Detection limits are often quoted as 0.1–1.0 % atomic percent (0.1 % = 1 part per thousand = 1000 ppm) for practical analyses, but lower limits may be achieved in many circumstances.

Measured Area

Measured area depends on instrument design. The minimum analysis area ranges from 10 to 200 micrometres. Largest size for a monochromatic beam of X-rays is 1–5 mm. Non-monochromatic beams are 10–50 mm in diameter. Spectroscopic image resolution levels of 200 nm or below has been achieved on latest imaging XPS instruments using synchrotron radiation as X-ray source.

Sample Size Limits

Instruments accept small (mm range) and large samples (cm range), e.g. wafers. Limiting factor is the design of the sample holder, the sample transfer, and the size of the vacuum chamber. Large samples are laterally moved in x and y direction to analyse a larger area.

Degradation during Analysis

- Depends on the sensitivity of the material to the wavelength of X-rays used, the total dose of the X-rays, the temperature of the surface and the level of the vacuum. Metals, alloys, ceramics and most glasses are not measurably degraded by either non-monochromatic or monochromatic X-rays. Some, but not all, polymers, catalysts, certain highly oxygenated compounds, various inorganic compounds and fine organics are degraded by either monochromatic or non-monochromatic X-ray sources.

- Non-monochromatic X-ray sources produce a significant amount of high energy Bremsstrahlung X-rays (1–15 keV of energy) which directly degrade the surface chemistry of various materials. Non-monochromatic X-ray sources also produce a significant amount of heat (100 to 200 °C) on the surface of the sample because the anode that produces the X-rays is typically only 1 to 5 cm (2 in) away from the sample. This level of heat, when combined with the Bremsstrahlung X-rays, acts synergistically to increase the amount and rate of degradation for certain materials. Monochromatic X-ray sources, because they are far away (50–100 cm) from the sample, do not produce any heat effects.

- Monochromatic X-ray sources are monochromatic because the quartz monochromator system diffracts the Bremsstrahlung X-rays out of the X-ray beam, which means the sample is only exposed to one narrow band of X-ray energy. For example, if aluminum K-alpha X-rays are used, the intrinsic energy band has a FWHM of 0.43 eV, centered on 1,486.7 eV ($E/\Delta E$ = 3,457). If magnesium K-alpha X-rays are used, the intrinsic energy band has a FWHM of 0.36 eV, centered on 1,253.7 eV ($E/\Delta E$ = 3,483). These are the intrinsic X-ray line widths; the range of energies to which the sample is exposed depends on the quality and optimization of the X-ray monochromator.

- Because the vacuum removes various gases (e.g., O_2, CO) and liquids (e.g., water, alcohol, solvents, etc.) that were initially trapped within or on the surface of the sample, the chemistry and morphology of the surface will continue to change until the surface achieves a steady state. This type of degradation is sometimes difficult to detect.

Use

XPS is, in effect, a non-destructive technique that measures the surface chemistry of most any material, however non-dry, outgassing, radioactive or highly magnetic materials can pose serious challenges.

Data Processing

Peak Identification

The number of peaks produced by a single element varies from 1 to more than 20. Tables of binding energies (BEs) that identify the shell and spin-orbit of each peak produced by a given element are included with modern XPS instruments, and can be found in various handbooks and websites. Because these experimentally determined BEs are characteristic of specific elements, they can be directly used to identify experimentally measured peaks of a material with unknown elemental composition.

Before beginning the process of peak identification, the analyst must determine if the BEs of the unprocessed survey spectrum (0-1400 eV) have or have not been shifted due to a positive or negative surface charge. This is most often done by looking for two peaks that due to the presence of carbon and oxygen.

Charge Referencing Insulators

Charge referencing is needed when a sample suffers either a positive (+) or negative (-) charge induced shift of experimental BEs. Charge referencing is needed to obtain meaningful BEs from both wide-scan, high sensitivity (low energy resolution) survey spectra (0-1100 eV), and also narrow-scan, chemical state (high energy resolution) spectra.

Charge induced shifting causes experimentally measured BEs of XPS peaks to appear at BEs that are greater or smaller than true BEs. Charge referencing is performed by adding or subtracting a "Charge Correction Factor" to each of the experimentally measured BEs. In general, the BE of the hydrocarbon peak of the C (1s) XPS signal is used to charge reference (charge correct) all BEs obtained from non-conductive (insulating) samples or conductors that have been deliberately insulated from the sample mount.

Charge induced shifting is normally due to: a modest excess of low voltage (-1 to -20 eV) electrons attached to the surface, or a modest shortage of electrons (+1 to +15 eV) within the top 1-12 nm of the sample caused by the loss of photo-emitted electrons. The degree of charging depends on various factors. If, by chance, the charging of the surface is excessively positive, then the spectrum might appear as a series of rolling hills, not sharp peaks as shown in the example spectrum.

The C (1s) BE of the hydrocarbon species (moieties) of the "Adventitious" carbon that appears on all, air-exposed, conductive and semi-conductive materials is normally found between 284.5 eV and 285.5 eV. For convenience, the C (1s) of hydrocarbon moieties is defined to appear between 284.6 eV and 285.0 eV. A value of 284.8 eV has become popular in recent years. However, some recent reports indicate that 284.9 eV or 285.0 eV represents hydrocarbons attached on metals, not the natural native oxide. The 284.8 eV BE is routinely used as the "Reference BE" for charge referencing insulators. When the C (1s) BE is used for charge referencing, then the charge correction factor is the difference between 284.8 eV and the experimentally measured C (1s) BE of the hydrocarbon moieties.

When using a monochromatic XPS system together with a low voltage electron flood gun for charge compensation the experimental BEs of the C (1s) hydrocarbon peak is often 4-5 eV smaller than the reference BE value (284.8 eV). In this case, all experimental BEs appear at lower BEs than expected and need to be increased by adding a value ranging from 4 to 5 eV. Non-monochromatic XPS systems are not usually equipped with a low voltage electron flood gun so the BEs will normally appear at higher BEs than expected. It is normal to subtract a charge correction factor from all BEs produced by a non-monochromatic XPS system.

Conductive materials and most native oxides of conductors should never need charge referencing. Conductive materials should never be charge referenced unless the topmost layer of the sample has a thick non-conductive film.

Peak-fitting

The process of peak-fitting high energy resolution XPS spectra is still a mixture of art, science, knowledge and experience. The peak-fit process is affected by instrument design, instrument components, experimental settings (aka analysis conditions) and sample variables. Most instrument parameters are constant while others depend on the choice of experimental settings.

Before starting any peak-fit effort, the analyst performing the peak-fit needs to know if the topmost 15 nm of the sample is expected to be a homogeneous material or is expected to be a mixture of materials. If the top 15 nm is a homogeneous material with only very minor amounts of adventitious carbon and adsorbed gases, then the analyst can use theoretical peak area ratios to enhance the peak-fitting process.

Variables that affect or define peak-fit results include:

- FWHMs
- Chemical Shifts
- Peakshapes
- Instrument design factors
- Experimental settings
- Sample factors

Full width at Half Maximum (FWHM)

The full width at half maximum (FWHM) values are useful indicators of chemical state changes and physical influences. That is, broadening of a peak may indicate: a change in the number of chemical bonds contributing to a peak shape, a change in the sample condition (x-ray damage) and/or differential charging of the surface (localised differences in the charge state of the surface). However, the FWHM also depends on the detector, and can also increase due to the sample getting charged.

When using high energy resolution experiment settings on an XPS equipped with a monochromatic Al K-alpha X-ray source, the FWHM of the major XPS peaks range from 0.3 eV to 1.7 eV. The following is a simple summary of FWHM from major XPS signals:

- Main metal peaks (e.g. 1s, 2p3, 3d5, 4f7) from pure metals have FWHMs that range from 0.30 eV to 1.0 eV.
- Main metal peaks (e.g. 1s, 2p3, 3d5, 4f7) from binary metal oxides have FWHMs that range from 0.9 eV to 1.7 eV.
- The O (1s) peak from binary metal oxides have FWHMs that, in general, range from 1.0 eV to 1.4 eV.
- The C (1s) peak from adventitious hydrocarbons have FWHMs that, in general, range from 1.0 eV to 1.4 eV.

Chemical Shifts

Chemical shift values depend on the degree of electron bond polarization between nearest neighbor atoms. A specific chemical shift is the difference in BE values of one specific chemical state versus the BE of one form of the pure element, or of a particular agreed-upon chemical state of that element. Component peaks derived from peak-fitting a raw chemical state spectrum can be assigned to the presence of different chemical states within the sampling volume of the sample.

Peak Shapes

Depends on instrument parameters, experimental parameters and sample characteristics.

- Instrument Design Factors: FWHM and purity of X-rays used (monochromatic Al, non-monochromatic Mg, Synchrotron, Ag, Zr).

    Design of electron analyzer (CMA, HSA, retarding field).

- Experiment Settings: Settings of the electron analyzer (e.g. pass energy, step size).

- Sample Factors: Physical form of the sample (single crystal, polished, powder, corroded). Number of physical defects within the analysis volume (from Argon ion etching, from laser cleaning).

Advanced Instrumentation Aspects

Hemispherical Electron Energy Analyzer

A hemispherical electron energy analyzer is generally used for applications where a higher resolution is needed. An ideal hemispherical analyzer consists of two concentric hemispherical electrodes (inner and outer hemispheres) held at proper voltages. It is possible to demonstrate that in such a system, (i) the electrons are linearly dispersed along the direction connecting the entrance and the exit slit, depending on their kinetic energy, while (ii) electrons with the same energy are first-order focused. When two potentials, V_1 and V_2, are applied to the inner and outer hemispheres, respectively, the electric potential and field in the region between the two electrodes can be calculated by solving the Laplace equation:

$$V(r) = -\left[\frac{(V_2 - V_1)}{(R_2 - R_1)}\right] \cdot \frac{(R_1 R_2)}{r} + const.$$

$$|E(r)| = -\left[\frac{(V_2 - V_1)}{(R_2 - R_1)}\right] \cdot \frac{(R_1 R_2)}{r^2}$$

where R_1 and R_2 are the radii of the two hemispheres. In order for the electrons with kinetic energy Eo to follow a circular trajectory of radius $R_0 = \frac{(R1 + R2)}{2}$, the force exerted by the electric field ($F_E = -e|E(r)|$) must equal the centripetal force (F_C) along the whole path. After some algebra, the following expression can be derived for the potential:

$$V(r) = \left(\frac{V_0 R_0}{r}\right) + const.,$$

where $V_0 = \frac{E_0}{e}$ is the energy of the electrons expressed in eV. From this equation, we can calculate the potential difference between the two hemispheres, which is given by:

$$V_2 - V_1 = V_0 \left(\frac{R_2}{R_1} - \frac{R_1}{R_2} \right).$$

The latter equation can be used to determine the potentials to be applied to the hemispheres in order to select electrons with energy $E_0 = |e| V_0$, the so-called *pass energy*.

In fact, only the electrons with energy E_0 impinging normal to the entrance slit of the analyzer describe a trajectory of radius $R_0 = (R_1 + R_2)/2$ and reach the exit slit, where they are revealed by the detector.

The instrumental energy resolution of the device depends both on the geometrical parameters of the analyzer and on the angular divergence of the incoming photoelectrons:

$$\Delta E = E_0 \left(\frac{w}{2R_0} + \frac{\alpha^2}{4} \right),$$

where w is the average width of the two slits, and α is the incidence angle of the incoming photoelectrons. Though the resolution improves with increasing R_0, technical problems related to the size of the analyzer put a limit on the actual value of R_0. Although a low pass energy E_0 improves the resolution, the electron transmission probability is reduced at low pass energy, and the signal-to-noise ratio deteriorates, accordingly. The electrostatic lenses in front of the analyzer have two main purposes: They collect and focus the incoming photoelectrons into the entrance slit of the analyzer, and they decelerate the electrons to the kinetic energy E_0, in order to increase the resolution.

When acquiring spectra in sweep (or scanning) mode, the voltages of the two hemispheres V_1 and V_2 - and hence the pass energy- are held fixed; at the same time, the voltage applied to the electrostatic lenses is swept in such a way that each channel counts electrons with the selected kinetic energy for the selected amount of time. In order to reduce the acquisition time per spectrum, the so-called snapshot (or fixed) mode has been introduced. This mode exploits the relation between the kinetic energy of a photoelectron and its position inside the detector. If the detector energy range is wide enough, and if the photoemission signal collected from all the channels is sufficiently strong, the photoemission spectrum can be obtained in one single shot from the image of the detector.

Cylindrical Mirror Analyzer

Since the relevant information, in photoemission spectroscopy, is contained in the kinetic energy distribution of the photoelectrons, a specific device is needed to energy-filter the electrons emitted (or scattered) by the sample. Electrostatic monochromators are the most common choice. The older design, a CMA, represents a trade-off between the need for high count rates and high angular/energy resolution. The so-called cylindrical mirror analyzer (CMA) is mostly used for checking the elemental composition of the surface. It consists of two co-axial cylinders placed in front of the sample, the inner one being held at a positive potential, while the outer cylinder is held at a negative potential. Only the electrons with the right energy can pass through this set-up and are detected at the end. The count rates are high but the resolution (both in energy and angle) is poor.

Synchrotron based XPS

A breakthrough has been actually brought about in the last decades by the development of large scale synchrotron radiation facilities. Here, bunches of relativistic electrons kept on a circular orbit inside a storage ring are accelerated through bending magnets or insertion devices like wigglers and undulators to produce a high brilliance and high flux photon beam. The main advantages of using synchrotron light are:

- The high brilliance of this kind of radiation, which is orders of magnitude more intense and better collimated than the one produced by anode-based sources;

- The tunability of synchrotron radiation over a wide frequency range;

- Its high polarization;

- The high photon flux;

- The possibility of producing extremely short pulses at a frequency as high as a MHz.

The highest spectral brightness and narrowest beam energy dispersion is attained by undulators, which consist of periodic array of dipole magnets in which the electrons are forced to wiggle and thus to emit coherent light. Besides the high intensity, energy tunability is one of the most important advantages of synchrotron light compared to the light produced by conventional X-ray sources. In fact, a wide energy range (from the IR to the Hard X-ray region, depending on the energy of the electron bunch) is accessible by changing the undulator gaps between the arrays. Continuous energy spectra available from a synchrotron radiation source allows selection of photon energies yielding optimum photoionization cross-sections appropriate for probing a particular core level. The high photon flux, in addition, makes it possible to perform XPS experiments also from low density atomic species, such as molecular and atomic adsorbates.

Electron Detectors

Older Style Electron Detector

Electrons can be detected using an electron multiplier, usually a channeltron. This device essentially consists of a glass tub with a resistive coating on the inside. A high voltage is applied between the front and the end. An electron which enters the channeltron is accelerated to the wall, where it removes more electrons, in such a way that an electron avalanche is created, until a measurable current pulse is obtained.

Theoretical Aspects

Quantum Mechanical Treatment

When a photoemission event takes place, the following energy conservation rule holds:

$$h\nu = \mid E_b^\nu \mid + E_{kin}$$

where $h\nu$ is the photon energy, $|E_b^v|$ is the electron BE (with respect to the vacuum level) prior to ionization, and E_{kin} is the kinetic energy of the photoelectron. If reference is taken with respect to the Fermi level (as it is typically done in photoelectron spectroscopy) $|E_b^v|$ must be replaced by the sum of the binding energy (BE) relative to the Fermi level, $|E_b^F|$, and the sample work function, Φ_0.

From the theoretical point of view, the photoemission process from a solid can be described with a semiclassical approach, where the electromagnetic field is still treated classically, while a quantum-mechanical description is used for matter. The one—particle Hamiltonian for an electron subjected to an electromagnetic field is given by:

$$i\hbar\frac{\partial\psi}{\partial t} = \left[\frac{1}{2m}\left(\hat{p}-\frac{e}{c}\hat{A}\right)^2+\hat{V}\right]\psi = \hat{H}\psi,$$

where ψ is the electron wave function, A is the vector potential of the electromagnetic field and V is the unperturbed potential of the solid. In the Coulomb gauge ($\nabla\cdot A=0$), the vector potential commutes with the momentum operator ($[\hat{p},\hat{A}]=0$), so that the expression in brackets in the Hamiltonian simplifies to:

$$\left(\hat{p}-\frac{e}{c}\hat{A}\right)^2 = \hat{p}^2 - 2\frac{e}{c}\hat{A}\cdot\hat{p}+\left(\frac{e}{c}\right)^2\hat{A}^2$$

Actually, neglecting the $\nabla\cdot A$ term in the Hamiltonian, we are disregarding possible photocurrent contributions. Such effects are generally negligible in the bulk, but may become important at the surface. The quadratic term in A can be instead safely neglected, since its contribution in a typical photoemission experiment is about one order of magnitude smaller than that of the first term.

In first-order perturbation approach, the one-electron Hamiltonian can be split into two terms, an unperturbed Hamiltonian $\hat{H}_0$, plus an interaction Hamiltonian $\hat{H}'$, which describes the effects of the electromagnetic field:

$$\hat{H}' = -\frac{e}{mc}\hat{A}\times\hat{p}$$

In time-dependent perturbation theory, for an harmonic or constant perturbation, the transition rate between the initial state ψ_i and the final state ψ_f is expressed by Fermi's Golden Rule:

$$\frac{d\omega}{dt} \propto \frac{2\pi}{\hbar}|\langle\psi_f|\hat{H}'|\psi_i\rangle|^2\,\delta(E_f-E_i-h\nu),$$

where E_i and E_f are the eigenvalues of the unperturbed Hamiltonian in the initial and final state, respectively, and $h\nu$ is the photon energy. Fermi's Golden Rule uses the approximation that the perturbation acts on the system for an infinite time. This approximation is valid when the time that the perturbation acts on the system is much larger than the time needed for the transition.

It should be understood that this equation needs to be integrated with the density of states $\rho(E)$ which gives:

$$\frac{d\omega}{dt} \propto \frac{2\pi}{\hbar} |\langle \psi_f | \hat{H}' | \psi_i \rangle|^2 \, \rho(E_f) = |M_{fi}|^2 \, \rho(E_f)$$

In a real photoemission experiment the ground state core electron BE cannot be directly probed, because the measured BE incorporates both initial state and final state effects, and the spectral linewidth is broadened owing to the finite core-hole lifetime (τ).

Assuming an exponential decay probability for the core hole in the time domain ($\propto \exp{-t/\tau}$), the spectral function will have a Lorentzian shape, with a FWHM (Full Width at Half Maximum) Γ given by:

$$I_L(E) = \frac{I_0}{\pi} \frac{\Gamma/2}{(E - E_b)^2 + (\Gamma/2)^2}$$

From the theory of Fourier transforms, Γ and τ are linked by the indeterminacy relation:

$$\Gamma \tau \geq \hbar$$

The photoemission event leaves the atom in a highly excited core ionized state, from which it can decay radiatively (fluorescence) or non-radiatively (typically by *Auger* decay). Besides Lorentzian broadening, photoemission spectra are also affected by a Gaussian broadening, whose contribution can be expressed by,

$$I_G(E) = \frac{I_0}{\sigma\sqrt{2}} \exp\left(-\frac{(E - E_b)^2}{2\sigma^2} \right)$$

Three main factors enter the Gaussian broadening of the spectra: the experimental energy resolution, vibrational and inhomogeneous broadening. The first effect is caused by the non perfect monochromaticity of the photon beam which results in a finite bandwidth- and by the limited resolving power of the analyzer. The vibrational component is produced by the excitation of low energy vibrational modes both in the initial and in the final state. Finally, inhomogeneous broadening can originate from the presence of unresolved core level components in the spectrum.

Theory of Core Level Photoemission of Electrons

In a solid, also in elastic scattering events contribute to the photoemission process, generating electron-hole pairs which show up as an inelastic tail on the high BE side of the main photoemission peak. In some cases, we observe also energy loss features due to plasmon excitations. This can either a final state effect caused by core hole decay, which generates quantized electron wave excitations in the solid (intrinsic plasmons), or it can be due to excitations induced by photoelectrons travelling from the emitter to the surface (extrinsic plasmons). Due to the reduced coordination number of first-layer atoms, the plasma. Frequency of bulk and surface atoms are related by the following equation: $\omega_{surf} = \dfrac{\omega_{bulk}}{\sqrt{2}}$, so that surface and bulk plasmons can be easily distinguished from each other. Plasmon states in a solid are typically localized at the surface, and can strongly affect the electron Inelastic Mean Free Path (IMFP).

Vibrational Effects

Temperature-dependent atomic lattice vibrations, or phonons, can broaden the core level components and attenuate the interference patterns in an XPD (X-Ray Photoelectron Diffraction) experiment. The simplest way to account for vibrational effects is by multiplying the scattered single-photoelectron wave function ϕ_j by the Debye-Waller factor:

$$W_j = \exp\left(-\Delta k_j^2 \overline{U}_j^2\right),$$

where Δk_j^2 is the squared magnitude of the wave vector variation caused by scattering, and $\overline{U}_j^2$ is the temperature-dependent one-dimensional vibrational mean squared displacement of the jth emitter. In the Debye model, the mean squared displacement is calculated in terms of the Debye temperature, Θ_D, as:

$$\overline{U}_j^2(T) = 9\hbar^2 T^2 / m k_B \Theta_D.$$

Ultraviolet Photoelectron Spectroscopy

Ultraviolet Photoelectron Spectroscopy (UPS) operates on the same principles as XPS, the only difference being that ionising radiation at energies of 10s of eV are used to induce the photoelectric effect, as opposed to photons of greater than 1keV that are used in XPS. In the laboratory setting ultraviolet photons are produced using a gas discharge lamp, typically filled with helium, although other gases such as argon and neon can also be used. The photons emitted by helium gas have energies of 21.2eV (He I) and 40.8eV (He II).

As lower energy photons are used, most core level photoemissions are not accessible using UPS, so spectral acquisition is limited to the valence band region. There are two types of experiment performed using UPS: Valence band acquisition and electronic workfunction measurement.

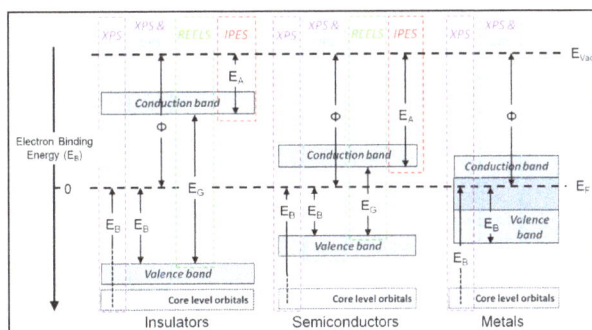

Valence Band

Many of the molecular orbitals from which valence band photoelectron signal originates posses a high degree of hybridisation, therefore the shifts in peak binding energy are far more varied and subtle than those observed for core level photoemission peaks. For this reason valence band spectra are predominantly used for material characterisation through spectral fingerprinting, and individual peak assignment is either performed on surfaces with well known electronic structure, or in conjunction with computational studies. Due to this ambiguity in the assignment of valence band peaks, these spectra are not used for quantification.

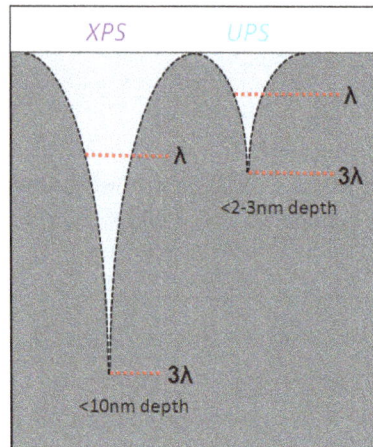

Difference in information depth for XPS and UPS.

XPS is also widely used to collect valence band spectra, the combination of both XPS and UPS to investigate the valence band can be extremely powerful as the ionisation cross section of an orbital is dependent on the incident photon energy, therefore different electronic transitions and states can be probed by using different photon energies.

UPS also exhibits greater surface sensitivity than XPS, the inherent surface sensitivity of XPS is due to the short inelastic mean free path (IMFP, or λ) of free electrons within a solid, with the so-called 'information depth' from which > 99% of a photoemission signal originates conventionally being defined at 3 mean free path lengths from the surface, which in XPS is often quoted as 10 nm. This is an approximation as the IMFP of an electron is determined by the material properties of the solid media through which it is travelling and its kinetic energy, with electrons of lower kinetic energy having shorter path lengths. The lower incident photon energies used in UPS give emit photoelectrons of much lower kinetic energies than those measured in XPS, therefore giving UPS an approximate information depth of 2-3nm.

Workfunction

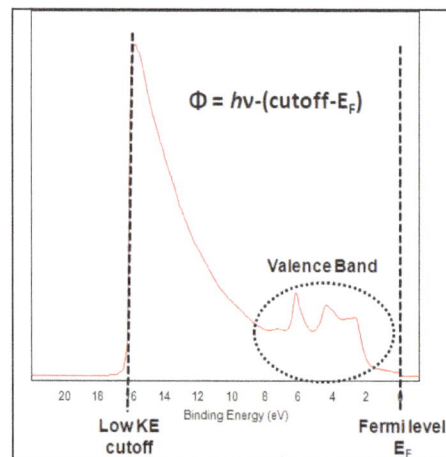

Measuring a material's work function using a UP spectrum.

The difference between the Fermi level and Vacuum level is referred to as the electronic work-function, a material property applied in the development of electronics devices, for example where matching of valence and conduction bands in multilayered devices is required. As a surface property, the workfunction is strongly influenced by variation in composition or structure at the surface, such as atmospheric contamination.

The electronic workfunction is acquired spectroscopically by measuring the difference between the Fermi Level and the cutoff of the 'tail' at the low kinetic energy end of the spectrum (a.k.a. spectrum width) and subtracting this value from the incident photon energy. This value can of course be measured using X-ray incident radiation, however UPS allows workfunction calculation from a single spectrum.

When measuring the electronic workfunction using photoelectron spectroscopy it is necessary to apply a small bias (typically 5-10 V) to the sample surface, so as to deconvolute the true workfunction of the surface from the internal workfunction of the spectrometer.

Auger Electron Spectroscopy

Auger electron spectroscopy (AES) is a nondestructive core-level electron spectroscopy for semi-quantitative determination of the elemental composition of surfaces, thin films, and interfaces. Auger electron spectroscopy (AES) identifies elemental compositions of surfaces by measuring the energies of Auger electrons. Auger electron emission is stimulated by bombarding the sample with an electron beam. The Auger electron energies are characteristic of the elements from which the electrons come.

It is based on the measurement of the kinetic energy emitted by the Auger electrons of an excited atom on the surface irradiated with a high-energy electron beam.

The energy emitted when the second or Auger electron transfers to another orbit or level is characteristic of the atom that it is a part of. The kinetic energy is determined with a scanning Auger microscope (SAM), a powerful tool that can analyze all elements on a solid surface, except for hydrogen and helium.

Auger electron spectroscopy is effective for analyzing the composition as well as the estimated concentration of the elements on the surface of solid materials. It is based on the Auger effect, a phenomenon that was observed and reported independently by Lise Meitner and Pierre Victor Auger in 1923 and 1925 respectively.

Some of the important analytical requirements are: an ultrahigh vacuum (UHV) environment; solid, conducting and semi-conducting surface samples; and special procedures for non-conducting or insulating samples.

Electron Transitions and the Auger Effect

The Auger effect is an electronic process at the heart of AES resulting from the inter- and intrastate transitions of electrons in an excited atom. When an atom is probed by an external mechanism, such as a photon or a beam of electrons with energies in the range of several eV to 50 keV, a core state electron can be removed leaving behind a hole. As this is an unstable state, the core hole can be filled by an outer shell electron, whereby the electron moving to the lower energy level loses an amount of energy equal to the difference in orbital energies. The transition energy can be coupled to a second outer shell electron, which will be emitted from the atom if the transferred energy is greater than the orbital binding energy. An emitted electron will have a kinetic energy of:

$$E_{\text{kin}} = E_{\text{Core State}} - E_B - E_C'$$

where $E_{\text{Core State}}$, E_B, E_C' are respectively the core level, first outer shell, and second outer shell electron binding energies (measured from the vacuum level) which are taken to be positive. The apostrophe (tic) denotes a slight modification to the binding energy of the outer shell electrons due to the ionized nature of the atom; often however, this energy modification is ignored in order to ease calculations. Since orbital energies are unique to an atom of a specific element, analysis of the ejected electrons can yield information about the chemical composition of a surface. The figure below illustrates two schematic views of the Auger process.

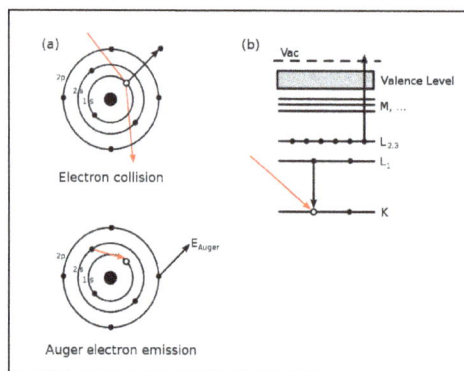

The figure above shows two views of the Auger process. (a) illustrates sequentially the steps involved in Auger deexcitation. An incident electron creates a core hole in the 1s level. An electron from the 2s level fills in the 1s hole and the transition energy is imparted to a 2p electron that is emitted. The final atomic state thus has two holes, one in the 2s orbital and the other in the 2p orbital. (b) illustrates the same process using X-ray notation.

The types of state-to-state transitions available to electrons during an Auger event are dependent on several factors, ranging from initial excitation energy to relative interaction rates, yet are often dominated by a few characteristic transitions. Because of the interaction between an electron's spin and orbital angular momentum (spin-orbit coupling) and the concomitant energy level splitting for various shells in an atom, there are a variety of transition pathways for filling a core hole. Energy levels are labeled using a number of different schemes such as the j-j coupling method for heavy elements ($Z \geq 75$), the Russell-Saunders L-S method for lighter elements ($Z < 20$), and a combination of both for intermediate elements. The j-j coupling method, which is historically linked to X-ray notation, is almost always used to denote Auger transitions. Thus for a $KL_1L_{2,3}$ transition, K represents the core level hole, L_1 the relaxing electron's initial state, and $L_{2,3}$ the emitted electron's initial energy state. Figure illustrates this transition with the corresponding spectroscopic notation. The energy level of the core hole will often determine which transition types will be favored. For single energy levels, i.e. K, transitions can occur from the L levels, giving rise to strong KLL type peaks in an Auger spectrum. Higher level transitions can also occur, but are less probable. For multi-level shells, transitions are available from higher energy orbitals (different n, ℓ quantum numbers) or energy levels within the same shell (same n, different ℓ number). The result are transitions of the type LMM and KLL along with faster Coster–Kronig transitions such as LLM. While Coster–Kronig transitions are faster, they are also less energetic and thus harder to locate on an Auger spectrum. As the atomic number Z increases, so too does the number of potential Auger transitions. Fortunately, the strongest electron-electron interactions are between levels that are close together, giving rise to characteristic peaks in an Auger spectrum. KLL and LMM peaks are some of the most commonly identified transitions during surface analysis. Finally, valence band electrons can also fill core holes or be emitted during KVV-type transitions.

Several models, both phenomenological and analytical, have been developed to describe the energetics of Auger transitions. One of the most tractable descriptions, put forth by Jenkins and Chung, estimates the energy of Auger transition ABC as:

$$E_{ABC} = E_A(Z) - 0.5[E_B(Z) + E_B(Z+1)] - 0.5[E_C(Z) + E_C(Z+1)]$$

$E_i(Z)$ are the binding energies of the ith level in element of atomic number Z and $E_i(Z+1)$ are the energies of the same levels in the next element up in the periodic table. While useful in practice, a more rigorous model accounting for effects such as screening and relaxation probabilities between energy levels gives the Auger energy as:

$$E_{ABC} = E_A - E_B - E_C - F(BC:x) + R_{xin} + R_{xex}$$

where $F(BC:x)$ is the energy of interaction between the B and C level holes in a final atomic state x and the R's represent intra- and extra-atomic transition energies accounting for electronic screening. Auger electron energies can be calculated based on measured values of the various E_i and compared to peaks in the secondary electron spectrum in order to identify chemical species. This technique has been used to compile several reference databases used for analysis in current AES setups.

Experimental Setup and Quantification

Instrumentation

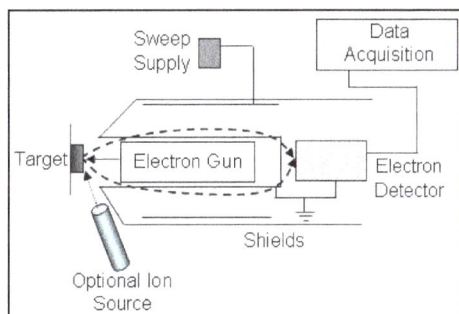

AES experimental setup using a cylindrical mirror analyzer (CMA). An electron beam is focused onto a specimen and emitted electrons are deflected around the electron gun and pass through an aperture towards the back of the CMA. These electrons are then directed into an electron multiplier for analysis. Varying voltage at the sweep supply allows derivative mode plotting of the Auger data. An optional ion gun can be integrated for depth profiling experiments.

Surface sensitivity in AES arises from the fact that emitted electrons usually have energies ranging from 50 eV to 3 keV and at these values, electrons have a short mean free path in a solid. The escape depth of electrons is therefore localized to within a few nanometers of the target surface, giving AES an extreme sensitivity to surface species. Because of the low energy of Auger electrons, most AES setups are run under ultra-high vacuum (UHV) conditions. Such measures prevent electron scattering off of residual gas atoms as well as the formation of a thin "gas (adsorbate) layer" on the surface of the specimen, which degrades analytical performance. A typical AES setup is shown schematically in figure. In this configuration, focused electrons are incident on a sample and emitted electrons are deflected into a cylindrical mirror analyzer (CMA). In the detection unit, Auger electrons are multiplied and the signal sent to data processing electronics. Collected Auger electrons are plotted as a function of energy against the broad secondary electron background spectrum.

Since the intensity of the Auger peaks may be small compared to the noise level of the background, AES is often run in a derivative mode that serves to highlight the peaks by modulating the electron collection current via a small applied AC voltage. Since this $\Delta V = k\sin(\omega t)$, the collection current becomes $I(V + k\sin(\omega t))$. Taylor expanding gives:

$$I(V + k\sin(\omega t)) \approx I_0 + I'(V + k\sin(\omega t)) + O(I'')$$

Using the setup in figure, detecting the signal at frequency ω will give a value for I' or $\dfrac{dN}{dE}$.

Plotting in derivative mode also emphasizes Auger fine structure, which appear as small secondary peaks surrounding the primary Auger peak. These secondary peaks, not to be confused with high energy satellites, arise from the presence of the same element in multiple different chemical states on a surface (i.e. Adsorbate layers) or from relaxation transitions involving valence band electrons of the substrate. Figure illustrates a derivative spectrum from a copper nitride film clearly showing the Auger peaks. The peak in derivative mode is not the true Auger peak, but rather the point of maximum slope of N(E), but this concern is usually ignored.

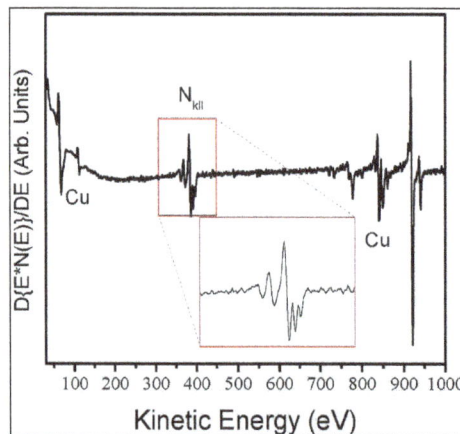

Figure: Auger spectrum of a copper nitride film in derivative mode plotted as a function of energy. Different peaks for Cu and N are apparent with the N KLL transition highlighted.

Quantitative Analysis

Semi-quantitative compositional and element analysis of a sample using AES is dependent on measuring the yield of Auger electrons during a probing event. Electron yield, in turn, depends on several critical parameters such as electron-impact cross-section and fluorescence yield. Since the Auger effect is not the only mechanism available for atomic relaxation, there is a competition between radiative and non-radiative decay processes to be the primary de-excitation pathway. The total transition rate, ω, is a sum of the non-radiative (Auger) and radiative (photon emission) processes. The Auger yield, ω_A, is thus related to the fluorescence (x-ray) yield, ω_X, by the relation,

$$\omega_A = 1 - \omega_X = 1 - \frac{W_X}{W_X + W_A}.$$

where W_X is the X-ray transition probability and W_A is the Auger transition probability. Attempts to relate the fluorescence and Auger yields to atomic number have resulted in plots similar to figure. A clear transition from electron to photon emission is evident in this chart for increasing atomic number. For heavier elements, x-ray yield becomes greater than Auger yield, indicating an increased difficulty in measuring the Auger peaks for large Z-values. Conversely, AES is sensitive to the lighter elements, and unlike X-ray fluorescence, Auger peaks can be detected for elements as light as lithium ($Z = 3$). Lithium represents the lower limit for AES sensitivity since the Auger effect is a "three state" event necessitating at least three electrons. Neither H nor He can be detected with this technique. For K-level based transitions, Auger effects are dominant for $Z < 15$ while for L- and M-level transitions, AES data can be measured for $Z \leq 50$. The yield limits effectively prescribe a cutoff for AES sensitivity, but complex techniques can be utilized to identify heavier elements, such as uranium and americium, using the Auger effect.

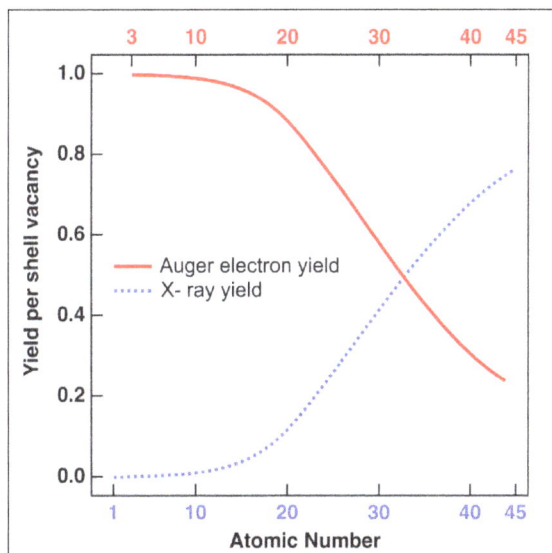

Fluorescence and Auger electron yields as a function of atomic number for K shell vacancies. Auger transitions (red curve) are more probable for lighter elements, while X-ray yield (dotted blue curve) becomes dominant at higher atomic numbers. Similar plots can be obtained for L and M shell transitions. Coster – Kronig (i.e. intra-shell) transitions are ignored in this analysis.

Another critical quantity that determines yield of Auger electrons at a detector is the electron impact cross-section. Early approximations (in cm²) of the cross-section were based on the work of Worthington and Tomlin,

$$\sigma_{ax}(E) = 1.3 \times 10^{13} b \frac{C}{E_p}$$

with b acting as a scaling factor between 0.25 and 0.35, and C a function of the primary electron beam energy, E_p. While this value of σ_{ax} is calculated for an isolated atom, a simple modification can be made to account for matrix effects:

$$\sigma(E) = \sigma_{ax}[1 + r_m(E_p, \alpha)]$$

where α is the angle to the surface normal of the incident electron beam; r_m can be established empirically and encompasses electron interactions with the matrix such as ionization due to backscattered electrons. Thus the total yield can be written as:

$$Y(t) = N_x \times \delta t \times \sigma(E,t)[1 - \omega_X] \exp\left(-t \cos \frac{\theta}{\lambda}\right) \times I(t) \times T \times \frac{d(\Omega)}{4\pi}.$$

Here N_x is the number of x atoms per volume, λ the electron escape depth, θ the analyzer angle, T the transmission of the analyzer, $I(t)$ the electron excitation flux at depth t, $d\Omega$ the solid angle, and δt is the thickness of the layer being probed. Encompassed in these terms, especially the Auger yield, which is related to the transition probability, is the quantum mechanical overlap of the initial and final state wave functions. Precise expressions for the transition probability, based on first-order perturbation Hamiltonians, can be found in Thompson and Baker. Often, all of these terms are not known, so most analyses compare measured yields with external standards of known composition. Ratios of the acquired data to standards can eliminate common terms, especially experimental setup characteristics and material parameters, and can be used to determine element composition. Comparison techniques work best for samples of homogeneous binary materials or uniform surface layers, while elemental identification is best obtained from comparison of pure samples.

Applications

AES has widespread use owing to its ability to analyze small spot sizes with diameters from 5 µm down to 10 nm depending on the electron gun. For instance, AES is commonly employed to study film growth and surface-chemical composition, as well as grain boundaries in metals and ceramics. It is also used for quality control surface analyses in integrated circuit production lines due to short acquisition times. Moreover, AES is used for areas that require high spatial resolution, which XPS cannot achieve. AES can also be used in conjunction with transmission electron microscopy (TEM) and scanning electron microscopy (SEM) to obtain a comprehensive understanding of microscale materials, both chemically and structurally. As an example of combining techniques to investigate microscale materials, figure shows the characterization of a single wire from a Sn-Nb multi-wire alloy. Figure is a SEM image of the singular wire and figure is a schematic depicting the distribution of Nb and Sn within the wire. Point analysis was performed along the length of the wire to determine the percent concentrations of Nb and Sn.

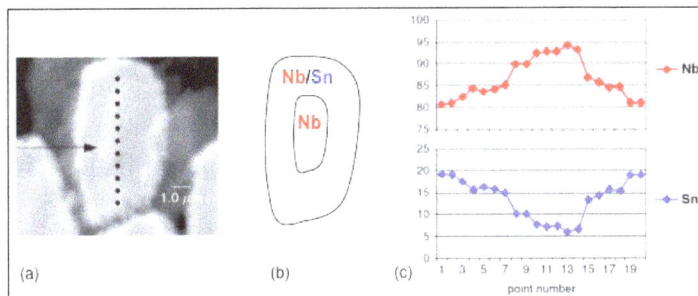

Figure: Analysis of a Sn-Nb wire. (a) SEM image of the wire, (b) schematic of the elemental distribution, and (c) graphical representation of point analysis giving the percent concentration of Nb and Sn.

AES is widely used for depth profiling. Depth profiling allows the elemental distributions of layered samples 0.2 – 1 μm thick to be characterized beyond the escape depth limit of an electron. Varying the incident and collection angles, and the primary beam energy controls the analysis depth. In general, the depth resolution decreases with the square root of the sample thickness. Notably, in AES, it is possible to simultaneously sputter and collect Auger data for depth profiling. The sputtering time indicates the depth and the intensity indicates elemental concentrations. Since, the sputtering process does not affect the ejection of the Auger electron, helium or argon ions can be used to sputter the surface and create the trench, while collecting Auger data at the same time. The depth profile does not have the problem of diffusion of hydrocarbons into the trenches. Thus, AES is better for depth profiles of reactive metals (e.g., gold or any metal or semiconductor). Yet, care should be taken because sputtering can mix up different elements, changing the sample composition.

Limitations

While AES is a very valuable surface analysis technique, there are limitations. Because AES is a three-electron process, elements with less than three electrons cannot be analyzed. Therefore, hydrogen and helium cannot be detected. Nonetheless, detection is better for lighter elements with fewer transitions. The numerous transition peaks in heavier elements can cause peak overlap, as can the increased peak width of higher energy transitions. Detection limits of AES include 0.1 – 1% of a monolayer, $10^{-16} – 10^{-15}$ g of material, and $10^{12} – 10^{13}$ atoms/cm^2.

Another limitation is sample destruction. Although focusing of the electron beam can improve resolution; the high-energy electrons can destroy the sample. To limit destruction, beam current densities of greater than 1 mA/cm^2 should be used. Furthermore, charging of the electron beam on insulating samples can deteriorate the sample and result in high-energy peak shifts or the appearance of large peaks.

Electron Energy Loss Spectroscopy

Electron energy loss spectroscopy (EELS) is the use of the energy distribution of electrons that pass through a thin sample to analyze the content of the sample and create images with unique contrast effects.

EELS instrumentation is typically incorporated into a transmission electron microscope (TEM) or a scanning TEM (STEM). These microscope types use high energy electrons (60 – 300 kV typically) to interrogate the sample. As the name implies, the electrons must "transmit" through the sample and thus requires an electron transparent sample. The electrons can interact either elastically (no energy exchange) or inelastically with the sample, and it is these interactions that EELS exploits to extract information about the sample.

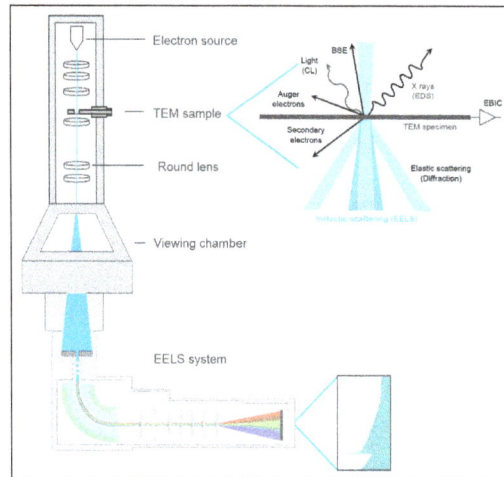

Specimen Information Provided by EELS signal

EELS data typically consists of either detailed, energy loss spectral information from the sample (spectroscopy) or images that have contrast created by the energy loss properties of the distribution of material in the sample (energy-filtered TEM). You can combine these methods in a technique called spectrum imaging, where spectral information is collected in a spatially resolved manner. This data is used to extract a wealth of information from the sample, including:

- A typical energy loss spectrum: The spectrum has many prominent features and is typically separated into three regions: electrons that have not lost energy (zero-loss peak), electrons that have interacted with the weakly bound electrons in the sample (low-loss distribution), and the electrons that have interacted with the tightly bound core electrons of the atoms:

  - Specimen thickness – Zero-loss peak (ZLP) and total spectrum intensity.

  - Valence/conduction electron density – Plasmon peaks.

  - Optical response (complex dielectric function) – Low-loss distribution.

  - Band structure and interband transitions – Near zero-loss features.

  - Elemental composition – Core-loss edges.

  - Bonding and oxidation state (density of unoccupied states) – Near edge fine structure (ELNES).

  - Distribution of near neighboring atoms (radial distribution function RDF) – Extended energy loss fine structure (ExELFS).

In addition to these spectrally related signals, you can create images by filtering in energy space (e.g., allows only the zero-loss electrons to form the image) can improve the contrast and resolution of the image and create unique image contrast.

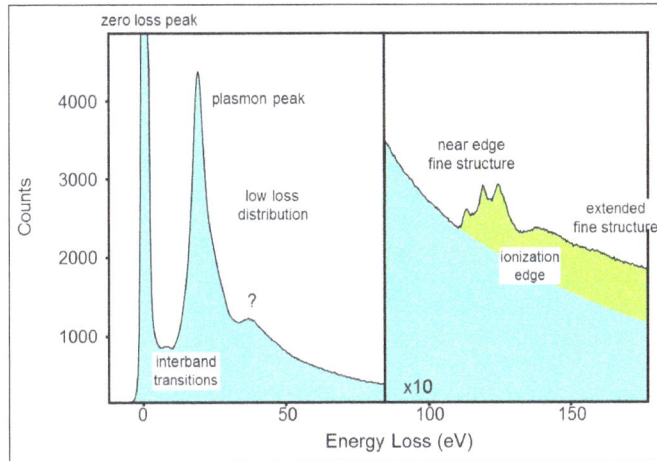

EELS and EDX

EELS is spoken of as being complementary to energy-dispersive x-ray spectroscopy (variously called EDX, EDS, XEDS, etc.), which is another common spectroscopy technique available on many electron microscopes. EDX excels at identifying the atomic composition of a material, is quite easy to use, and is particularly sensitive to heavier elements. EELS has historically been a more difficult technique but is in principle capable of measuring atomic composition, chemical bonding, valence and conduction band electronic properties, surface properties, and element-specific pair distance distribution functions. EELS tends to work best at relatively low atomic numbers, where the excitation edges tend to be sharp, well-defined, and at experimentally accessible energy losses (the signal being very weak beyond about 3 keV energy loss). EELS is perhaps best developed for the elements ranging from carbon through the 3d transition metals (from scandium to zinc). For carbon, an experienced spectroscopist can tell at a glance the differences between diamond, graphite, amorphous carbon, and "mineral" carbon (such as the carbon appearing in carbonates). The spectra of 3d transition metals can be analyzed to identify the oxidation states of the atoms. Cu(I), for instance, has a different so-called "white-line" intensity ratio than does Cu(II). This ability to "fingerprint" different forms of the same element is a strong advantage of EELS over EDX. The difference is mainly due to the difference in energy resolution between the two techniques (~1 eV or better for EELS, perhaps a few tens of eV for EDX).

Variants

There are several basic flavors of EELS, primarily classified by the geometry and by the kinetic energy of the incident electrons (typically measured in kiloelectron-volts, or keV). Probably the most common today is transmission EELS, in which the kinetic energies are typically 100 to 300 keV and the incident electrons pass entirely through the material sample. Usually this occurs in a transmission electron microscope (TEM), although some dedicated systems exist which enable extreme resolution in terms of energy and momentum transfer at the expense of spatial resolution.

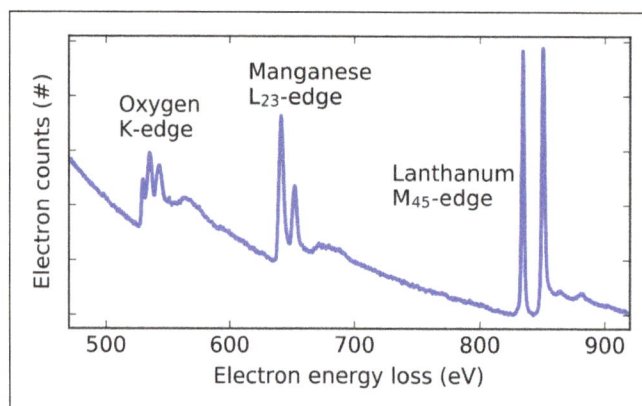

Example of inner shell ionization edge (core loss) EELS data from $La_{0.7}Sr_{0.3}MnO_3$, acquired on a scanning transmission electron microscope.

Other flavors include reflection EELS (including reflection high-energy electron energy-loss spectroscopy (RHEELS)), typically at 10 to 30 keV, and aloof EELS (sometimes called near-field EELS), in which the electron beam does not in fact strike the sample but instead interacts with it via the long-ranged Coulomb interaction. Aloof EELS is particularly sensitive to surface properties but is limited to very small energy losses such as those associated with surface plasmons or direct interband transitions.

Within transmission EELS, the technique is further subdivided into valence EELS (which measures plasmons and interband transitions) and inner-shell ionization EELS (which provides much the same information as x-ray absorption spectroscopy, but from much smaller volumes of material). The dividing line between the two, while somewhat ill-defined, is in the vicinity of 50 eV energy loss.

The Electron Energy Loss (EEL) Spectrum

The EEL spectrum can be roughly split into two different regions: the low-loss spectrum (up until about 50eV in energy loss) and the high-loss spectrum. The low-loss spectrum contains the zero-loss peak as well as the plasmon peaks, and contains information about the band structure and dielectric properties of the sample. The high-loss spectrum contains the ionisation edges that arise due to inner shell ionisations in the sample. These are characteristic to the species present in the sample, and as such can be used to obtain accurate information about the chemistry of a sample.

Thickness Measurements

EELS allows quick and reliable measurement of local thickness in transmission electron microscopy. The most efficient procedure is the following:

- Measure the energy loss spectrum in the energy range about −5.200 eV (wider better). Such measurement is quick (milliseconds) and thus can be applied to materials normally unstable under electron beams.

- Analyse the spectrum: (i) extract zero-loss peak (ZLP) using standard routines; (ii) calculate integrals under the ZLP (I_o) and under the whole spectrum (I).

- The thickness t is calculated as mfp*$ln(I/I_o)$. Here mfp is the mean free path of electron inelastic scattering, which has recently been tabulated for most elemental solids and oxides.

The spatial resolution of this procedure is limited by the plasmon localization and is about 1 nm, meaning that spatial thickness maps can be measured in scanning transmission electron microscopy with ~1 nm resolution.

Pressure Measurements

The intensity and position of low-energy EELS peaks are affected by pressure. This fact allows mapping local pressure with ~1 nm spatial resolution.

- Peak shift method is reliable and straightforward. The peak position is calibrated by independent (usually optical) measurement using a diamond anvil cell. However, the spectral resolution of most EEL spectrometers (0.3-2 eV, typically 1 eV) is often too crude for the small pressure-induced shifts. Therefore, the sensitivity and accuracy of this method is relatively poor. Nevertheless, pressures as small as 0.2 GPa inside helium bubbles in aluminum have been measured.

- Peak intensity method relies on pressure-induced change in the intensity of dipole-forbidden transitions. Because this intensity is zero for zero pressure the method is relatively sensitive and accurate. However, it requires existence of allowed and forbidden transitions of similar energies and thus is only applicable to specific systems, e.g., Xe bubbles in aluminum.

EELS in Confocal Geometry

Scanning confocal electron energy loss microscopy (SCEELM) is a new analytical microscopy tool that enables a double corrected transmission electron microscope to achieve sub-10 nm depth resolution in depth sectioning imaging of nanomaterials. It was previously termed as energy filtered scanning confocal electron microscopy due to the lack to full spectrum acquisition capability (only a small energy window on the order of 5 eV can be used at a time). SCEELM takes advantages of the newly developed chromatic aberration corrector which allows electrons of more than 100 eV of energy spread to be focused to roughly the same focal plane. It has been demonstrated that a simultaneous acquisition of the zero loss, low-loss, and core loss signals up to 400 eV in the confocal geometry with depth discrimination capability.

EELS Limitations

The main limitation of the EELS technique is the effect of sample thickness. As the sample thickness increases, the strong interaction of primary electrons within the sample results in the electrons undergoing multiple energy loss events. This tends to reduce the signal-to-background ratio of the EELS edges, reducing the visibility of the edges. However, since the spatial resolution of transmission electron microscope techniques is also reduced as the thickness increases, this loss of the edge visibility is not always the largest impediment to quality data.

Electronic Transitions

Like atoms, due to different configurations of electrons, molecules display several energy levels. They are characterized by spin (singlet, doublet, triplet.) and by an angular moment, or, more precisely, by L_z which is the projection of this moment parallel to the axis of the molecule which represents a special direction. The respective eigenvalues Λ are denoted with greek letters:

$$\Lambda \qquad \pm1 \quad \pm2 \quad \pm3 \quad \pm4 \quad \pm5 \quad \ldots$$
$$\text{Trem symbol} \quad \Sigma \quad \Pi \quad \Delta \quad \Gamma \quad \Gamma \quad \ldots$$

In analogy with S, P, D, F, G, ... to quantum numbers l = 0, 1, 2, 3, 4, ... for atoms. As a first approximation, we describe the energy of a diatomic molecule as a sum of electronic, vibrational and rotational energies.

$$E = E_{el} + E_{vib} + E_{rot}$$
$$E = T_e + \omega_e \left(v + \tfrac{1}{2} \right) + BJ\left(J+1 \right),$$

where E_{el} is the minimum of the potential curve. With an electronic transition taking place, all three addends may change. Note that ω_e and B for the initial and final electronic state differ. The state of lower energy is doubly primed (v", J"), the one of higher energy primed (v', J'). A transition corresponds with a difference in energy ΔE.

$$\Delta E_{el} = E'_{el} - E''_{el} = T'_e - T''_e$$
$$\Delta E = E_{el} + \Delta E_{vib} + \Delta E_{rot} \qquad \Delta E_{vib} = \omega'_e \left(v' + \tfrac{1}{2} \right) - \omega''_e \left(v'' + \tfrac{1}{2} \right)$$
$$\Delta E_{rot} = B'J'\left(J'+1 \right) - B''J''\left(J''+1 \right)$$

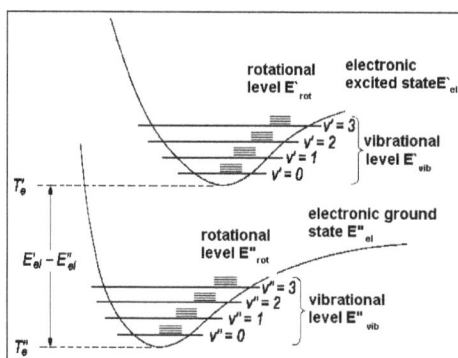

Selection rules for transitions have been established. The table below contains the criteria for transitions and lists examples.

	Allowed $\Delta\Lambda = 0, \pm 1$ and $\Delta S = 0$	Forbidden
	$^1\Pi \leftarrow {}^1\Sigma$	$^3\Pi \leftarrow {}^1\Sigma$, since $\Delta S = 1$
Example	$^1\Pi \leftarrow {}^1\Pi$	$^1\Pi \leftarrow {}^1\Phi$, since $\Delta\Lambda = -2$
	$^3\Delta \leftarrow {}^3\Delta$	$^3\Delta \leftarrow {}^1\Sigma$, since $\Delta S = 1$ and $\Delta\Lambda = 2$

If we consider the rotational states as well, it is required that the total angular momentum of photon and molecule remains constant. The selection rules depend on whether there are solely sigma states or whether an initial, a final state or both states have an electronic angular momentum J. The second case is referred to as general in the table below.

	$\Delta J = -1$	P branch
for $\Sigma \leftrightarrow \Sigma$	$\Delta J = +1$	R branch
	$\Delta J = -1$	P branch
in general	$\Delta J = 0$	Q branch
	$\Delta J = +1$	R branch

I.e., $\Delta J = 0$ is allowed since, within the system, there is at least one state with an electronic angular momentum that increases or decreases during transition.

No selection rules exist for changes of the vibrational state. This is attributed to remarkable differences in equilibrium distance between r_e' versus r_e'' and frequency ω_e' versus ω_e''. In terms of quantum mechanics, the initial and final potential curves differ. As a consequence, we deal with two wavefunctions Ψ_v and obtain different series of vibrational states v' and v" and the respective eigenvalues. Apart from this, the Franck-Condon-principle allows to calculate the probability of a transistion to some vibrational level of an excited state, i.e. the intensity of lines that reflect such transitions. An overlap of the following kind is the base to determine the probability as a function of v' and v".

$$\left| \int \Psi_{v'}{}^* \cdot \Psi_{v''} \cdot d\tau \right|^2$$

The obtained values have been named Franck-Condon-Factors.

$$v_0 = E_{el}' - E_{el}'' + \omega_e'\left(v' + \tfrac{1}{2}\right) - \omega_e''\left(v'' + \tfrac{1}{2}\right)$$

If we choose a transition $v' \leftarrow v''$, we expect the wavenumbers for the peaks within the P-, R- und Q-branch as given in the table below. The ratio of intensities for the three branches are decribed by the Hoenl-London-Factors.

$\Delta J = +1$	R branch	$v_R = v_0 + 2B_{v'} + (3B_{v'} - B_{v''})J'' - (B_{v''} - B_{v'})J''^2$	J" = 0, 1, 2, ...
$\Delta J = -1$	P branch	$v_P = v_0 - \left(B_{v'} + B_{v''}\right)J'' - \left(B_{v''} - B_{v'}\right)J''^2$	J" = 1, 2, ...
$\Delta J = 0$	Q branch	$v_Q = v_0 - \left(B_{v''} - B_{v'}\right)J'' - \left(B_{v''} - B_{v'}\right)J''^2$	J" = 1, 2, ...

Now we want to find out in which way an increase in the rotational quantum number J" affects the position of lines. To get a comprehensive picture that is founded on data from all three branches of the transition, we modify the equations above.

Branch	Substitution	Range
R	J" = z-1	for J" $\geq$ 0
P	J" = -z	for J" > 1

A substitution of quantum number J" by some counting variable z yields one equation that expresses the dependency between changes of the rotational state and the wavenumber ω for the transistion.

This substitution and rearrangement of the addends yield the equations:

$$v_R = v_0 + \left(B_{v'} + B_{v''}\right)z - \left(B_{v''} - B_{v'}\right)z^2 \quad z = 1, 2, 3, \dots$$
$$v_P = v_0 + \left(B_{v'} + B_{v''}\right)z - \left(B_{v''} - B_{v'}\right)z^2 \quad z = -1, -2, -3, \dots$$

Note that we recieved one equation v(z). We further introduce $\Delta B = B_{v''} - B_{v'}$ and $B = \frac{1}{2}(B_{v''} + B_{v'})$ as abbreviations and get one expression for the spectral position of transistions with $\Delta J = \pm 1$.

$$v = v_0 + 2Bz - \Delta Bz^2$$

An analogous derivation with a quasi-substition J" = z yields a second equation $v = v_0 - \Delta Bz - \Delta Bz^2$ for lines of the Q branch. Usually we have $|\Delta B| < B$, thus the lines of this branch accumulate close to v_0. If we plot a graph for pairs of v(z) and z using the horizontal axis for the wavenumbers, we obtain the so-called Fortrat parabolas.

a) Red shadowed Prevalent case, electronic excitation increases bond length.

$\Delta B > 0$
$r_e' > r_e''$
$z_{head} > 0$
$v_{head} > v_0$

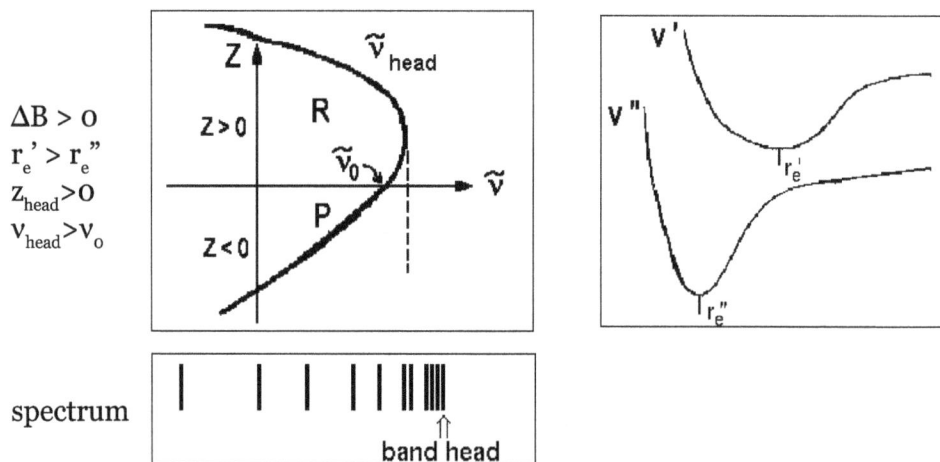

spectrum

band head

b) Blue shadowed in some rare cases, electronic excitation decreases bond length.

$\Delta B < 0$
$r_e' < r_e''$
$z_{head} < 0$
$v_{head} < v_0$

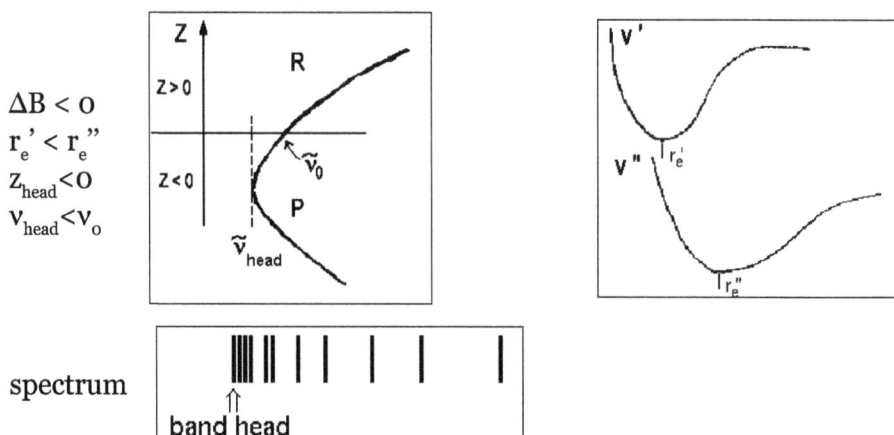

spectrum

band head

In case a) the maximal wavenumber/frequency is found for high rotations of the R branch, in case b) the lowest wavenumber/frequency is found for the highest rotations of the P branch. The frequency v_0 for z=0 is the so-called band origin. The minima and maxima respectively are the so-called band head. Mathematically, they are determined as the zeros of function dv(z)/dz.

$$\frac{d\tilde{v}}{dz} = 0 = 2B - 2\Delta B z_{head} \implies z_{head} = \frac{B}{\Delta B} \text{ and } v_{head} = v_0 + \frac{B^2}{\Delta B}$$

Franck-Condon Principle

The Franck-Condon Principle describes the intensities of vibronic transitions, or the absorption or emission of a photon. It states that when a molecule is undergoing an electronic transition, such as ionization, the nuclear configuration of the molecule experiences no significant change. This is due in fact that nuclei are much more massive than electrons and the electronic transition takes place faster than the nuclei can respond. When the nucleus realigns itself with with the new electronic configuration, the theory states that it must undergo a vibration.

If we picture the vertical transition from ground to excited electronic state as occurring from a vibrational wave function that gives a probability distribution of finding the nuclei in a give region of space we can determine the probability of a given vibrational level from the overlap integral Sv',v which gives the overlap of the vibrational wave function in the ground and excited state. The v' quantum numbers refer to the ground state and the vv quantum numbers refer to the excited state. The transition probability can be separated into electronic and nuclear parts using the Condon approximation.

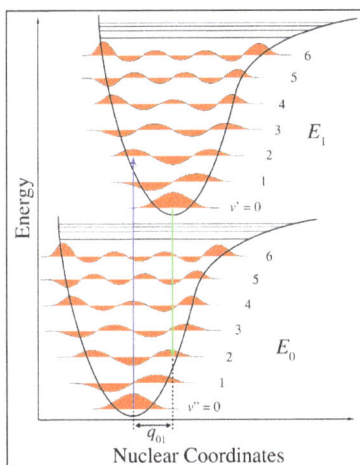

Figure: Franck–Condon principle energy diagram.

Since electronic transitions are very fast compared with nuclear motions, vibrational levels are favored when they correspond to a minimal change in the nuclear coordinates. The potential wells are shown favoring transitions between v=0v=0 and v'=2v'=2.

In figure above the nuclear axis shows a consequence of the internuclear separation and the vibronic transition is indicated by the blue and green vertical arrows. This figure demonstrates three things:

1. An absorption leads to a higher energy state,

2. Fluorescence leads to a lower energy state, and

3. The shift in nuclear coordinates between the ground and excited state is indicative of a new equilibrium position for nuclear interaction potential. The fact that the fluorescence arrow is shorter than the absorption indicates that it has less energy, or that its wavelength is longer.

The Classical Condon Approximation

Condon approximation is the assumption that the electronic transition occurs on a time scale short compared to nuclear motion so that the transition probability can be calculated at a fixed nuclear position.

This change in vibration is maintained during a state termed the rapid electronic excitation. The resulting Coulombic forces produce an equilibrium as shown in the figure for the nuclei termed a turning point. The turning point can be mapped by drawing a vertical line from the minimum of the lower curve to the intersection of the higher electronic state. This procedure is termed a vertical transition and was discussed before in the context of photoelectron spectroscopy (another electronic spectroscoy).

The Franck-Condon Principle explains the relative intensities of vibronic transitions by relating the probablity of a vibrational transition to the overlap of the vibrational wave functions. It states that the probability of a vibrational transition occurring is weighted by the Franck-Condon overlap integral:

$$P_{i \to f} = \left| \left\langle \psi^*_{final} \mid \mu \mid \psi_{initial} \right\rangle \right|^2 = \left| \int \psi^*_{final} \mu \psi_{initial} d\tau \right|^2$$

Within the Franck-Condon approximation, the nuclei are considered "fixed" during electronic transitions. Thus, electronic transitions can be considered vertical transitions on electronic potential energy curves.

The Quantum Franck-Condon Principle

The Franck-Condon Principle has both a Classical and Quantum application. Classically, the Franck–Condon principle is the approximation that an electronic transition is most likely to occur without changes in the positions of the nuclei in the molecular entity and its environment. The resulting state is called a Franck–Condon state, and the transition involved, a vertical transition. The quantum mechanical formulation of this principle is that the intensity of a vibronic transition is proportional to the square of the overlap integral between the vibrational wavefunctions of the two states that are involved in the transition.

The Franck-Condon principle is based on the Born-Oppenheimer approximation, which allows separation of the electronic q and nuclear Q wave functions given the total wavefunction.

$$\left| \psi_{total}(Q, q) \right\rangle = \left| \psi_{nuc}(Q) \right\rangle \psi el(Q; q) \right\rangle$$

Since the transition operator, $\hat{\mu}(q)$, is dependent only on the electronic component, the nuclear components can be separated from the transition moment integral that dictates the probability of the transition occuring:

$$\left\langle \psi^*_{total,f} \mid \hat{\mu} \mid \psi_{total,i} \right\rangle = \left\langle \psi^*_{nuc,f} \mid \left\langle \psi^*_{el,f} \mid \mu \mid \psi_{el,i} \right\rangle \mid \psi_{nuc,i} \right\rangle$$

$$= \underbrace{\left\langle \psi^*_{nuc,f} \mid \psi_{nuc,i} \right\rangle}_{\text{nuclear overlap}} \left\langle \psi^*_{el,f} \mid \mu \mid \psi_{el,i} \right\rangle.$$

If the nuclear overlap integral is zero for this transition, then the transition will not be observed, irrespective of the magnitude of the electronic factor.

S_{00} Transition Evaluated within Harmonic Oscillator Model

The nuclear overlap for the zero-zero transition S_{00} can be calculated quite simply using the definition of the Gaussian form of the harmonic oscillator wavefunctions.

The zero-point wavefunction in the ground electronic state is,

$$\mid \psi(R) \rangle = \left| \left(\frac{\alpha}{\pi} \right)^{1/4} e^{-\alpha(R-R_e)^2/2} \right\rangle$$

The zero-point wavefunction in the excited electronic state is,

$$\mid \psi(R) \rangle = \left| \left(\frac{\alpha}{\pi} \right)^{1/4} e^{-\alpha(R-Q_e)^2/2} \right\rangle$$

Where,

$$\alpha = \frac{\sqrt{mk}}{\hbar}$$

- R_e is the equilibrium bond length in the ground electronic state.
- Q_e is the equilibrium bond length in the excited electronic state.

The nuclear overlap integral is,

$$S_{00} = \left\langle \psi^*_{nuc,f} \mid \psi_{nuc,i} \right\rangle = \sqrt{\frac{\alpha}{\pi}} \int_\infty^\infty e^{-\alpha(R-R_e)^2/2} e^{-\alpha(R-Q_e)^2/2} dR$$

The exponent in Equation $S_{00} = \left\langle \psi^*_{nuc,f} \mid \psi_{nuc,i} \right\rangle = \sqrt{\frac{\alpha}{\pi}} \int_\infty^\infty e^{-\alpha(R-R_e)^2/2} e^{-\alpha(R-Q_e)^2/2} dR$ can be expanded as,

$$S_{00} = \sqrt{\frac{\alpha}{\pi}} \int_\infty^\infty e^{-\alpha(2R^2-RR_e-2RQ_e+R_e^2+Q_e^2)/2} dR$$

and we use,

$$(R_e + Q_e)^2 = R_e^2 + Q_e^2 + 2R_eQ_e$$

and

$$(R_e - Q_e)^2 = R_e^2 + Q_e^2 - 2R_eQ_e$$

to substitute and complete the square inside the integral. We can express,

$$R_e^2 + Q_e^2 = \frac{1}{2}[(R_e + Q_e)^2 + (R_e - Q_e)^2].$$

Thus, the integral in Equation $S_{00} = \sqrt{\frac{\alpha}{\pi}} \int_{\infty}^{\infty} e^{-\alpha(2R^2 - RR_e - 2RQ_e + R_e^2 + Q_e^2)/2} dR$ is

$$S_{00} = \sqrt{\frac{\alpha}{\pi}} e^{-\alpha(R_e - Q_e)^2/4} \int_{-\infty}^{\infty} e^{-\alpha\{R - 1/2(R_e + Q_e)\}^2} dR$$

The integral is a Gaussian integral. You can show that if we let $z = \sqrt{\alpha}\{R - 1/2(R_e + Q_e)\}$ then $dz = \sqrt{\alpha}dR$ and the integral becomes,

$$S_{00} = \sqrt{\frac{\alpha}{\pi}} e^{-\alpha(R_e - Q_e)^2/4} \frac{1}{\sqrt{\alpha}} \int_{-\infty}^{\infty} e^{z^2} dz$$

this integral has been solved already, from a table of integrals, the following equation

$$S_{00} = \sqrt{\frac{\alpha}{\pi}} e^{-\alpha(R_e - Q_e)^2/4} \frac{1}{\sqrt{\alpha}} \int_{-\infty}^{\infty} e^{z^2} dz \text{ becomes,}$$

$$S_{00} = e^{-\alpha(R_e - Q_e)^2/4}$$

We would follow the same procedure to calculate that overlap of the zeroth level vibration in the ground to the first excited vibrational level of the excited state: S_{01}.

S_{01} Transition Evaluated within Harmonic Oscillator Model

To calculate the overlap of zeroth ground state level ($v = 0$) with the first excited state level ($v' = 1$) we use the Hermite polynomial $H_1(x) = 2x$ for describing the excited state wavefunction. Here $x = \sqrt{\alpha}(R - Q_e)$.

$$S_{01} = \langle \psi_{nuc,f}^* | \psi_{nuc,i} \rangle$$

with the zero-point wavefunction in the ground electronic state is,

$$|\psi(R)\rangle = |\left(\frac{\alpha}{\pi}\right)^{1/4} e^{-\alpha(R - R_e)^2/2}\rangle$$

The first excited-state wavefunction in the excited electronic state is,

$$|\psi(R)\rangle = |\left(\frac{\alpha}{\pi}\right)^{1/4} \sqrt{\alpha}2(R - Q_e)e^{-\alpha(R - Q_e)^2/2}\rangle$$

The overlap of zeroth ground state level with the first excited state level (equation $S_{01} = \langle \psi^*_{nuc,f} | \psi_{nuc,i} \rangle$) is then,

$$S_{01} = \frac{1}{\sqrt{2}} \sqrt{\frac{\alpha}{\pi}} \int_{-\infty}^{\infty} e^{-\alpha(R-R_e)^2/2} \sqrt{\alpha}\, 2(R-Q_e) e^{-\alpha(R-Q_e)^2/2}$$

and

$$S_{01} = \sqrt{\frac{2\alpha^2}{\pi}} e^{-\alpha(R_e-Q_e)^2/4} \int_{-\infty}^{\infty} (R-Q_e) e^{-\alpha\{(R-1/2(Re+Qe)^2\}}$$

The same substitutions can be made as above so that the integral can be written as (not shown and to be demonstrated in a homework exercises) and the final result is,

$$S_{01} = \sqrt{\frac{\alpha^2}{2}} (R_e - Q_e) e^{-\alpha(R_e-Q_e)^2/4}$$

We could continue and calculate that overlap of the zeroth level in the ground state with all the higher light vibrational levels: S_{02}, S_{03}, etc. Each term corresponds to a transition with a different energy since the vibrational levels have different energies. The absorption band then has the appearance of a progression (a Franck-Condon progression) of transitions between different levels each with its own probability.

Franck-Condon Progressions

To understand the significance of the above formula for the FC factor, let us examine a ground and excited state potential energy surface at T=0 Kelvin. Shown below are two states separated by 8,000 cm^{-1} in energy. This is energy separation between the bottoms of their potential wells, but also between the respective zero-point energy levels. Let us assume that the wavenumber of the vibrational mode is 1,000 cm^{-1} and that the bond length is increased due to the fact that an electron is removed from a bonding orbital and placed in an anti-bonding orbital upon electronic excitation.

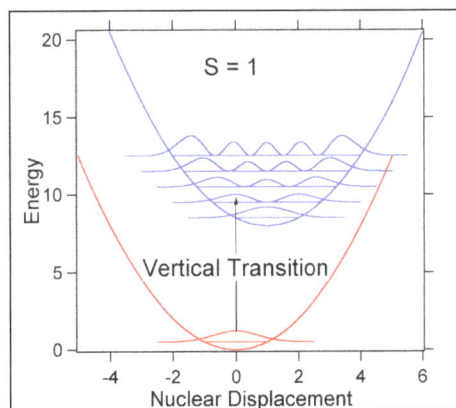

Figure: Wavefunctions transitions for a harmonic oscillator model system with moderate displacement (S=1).

According to the above model for the Franck-Condon factor we would generate a "stick" spectrum where each vibrational transition is infinitely narrow and transition can only occur when

$E = h\nu$ exactly. For example, the potential energy surfaces were given for S = 1 and the transition probability at each level is given by the sticks (black) in the figure below.

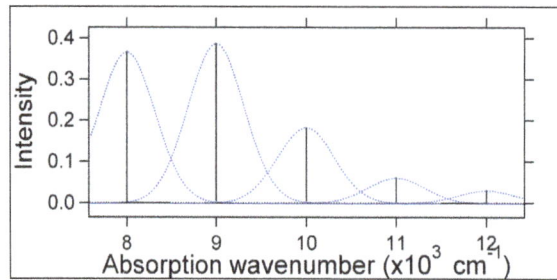

Figure: Stick spectrum, dressed with Gaussians, for the moderate displacement (S=1) harmonic oscillator system.

The dotted Gaussians that surround each stick give a more realistic picture of what the absorption spectrum should look like. In this first place each energy level (stick) will be given some width by the fact that the state has a finite lifetime. Such broadening is called homogeneous broadening since it affects all of the molecules in the ensemble in a similar fashion. There is also broadening due to small differences in the environment of each molecule. This type of broadening is called inhomogeneous broadening. Regardless of origin the model above was created using a Gaussian broadening.

The nuclear displacement between the ground and excited state determines the shape of the absorption spectrum. Let us examine both a smaller and a large excited state displacement. If $S = \frac{1}{2}$ and the potential energy surfaces in this case are:

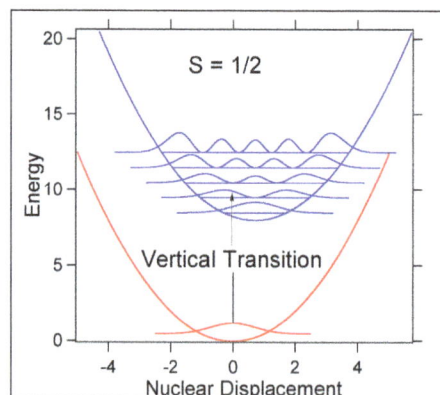

Figure: Wavefunctions transitions for a harmonic oscillator model system with small displacement (S=1/2).

For this case the "stick" spectrum has the appearance in figure below.

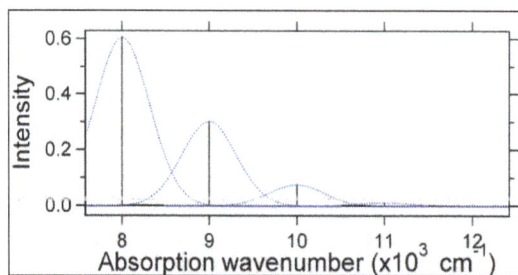

Figure: Stick spectrum, dressed with Gaussians, for the small displacement (S=1/2) harmonic oscillator system.

The zero-zero or $S_{0,0}$ vibrational transition is much large in the case where the displacement is small. As a general rule of thumb the S constant gives the ratio of the intensity of the $v = 2$ transition to the v=1 transition. In this case since S=0.5, the v=2 transition is 0.5 the intensity of v=1 transition.

As an example of a larger displacement the disposition of the potential energy surfaces for S = 2 is shown below.

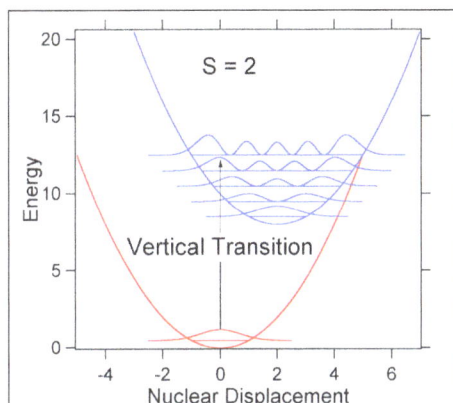

Figure: Wavefunctions transitions for a harmonic oscillator model system with strong displacement (S=2).

The larger displacement results in decreased overlap of the ground state level with the v = 0 level of the excited state. The maximum intensity will be achieved in higher vibrational levels as shown in the stick spectrum.

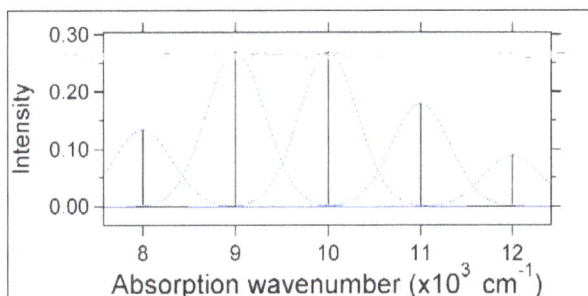

Figure: Stick spectrum, dressed with Gaussians, for the large displacement (S=2) harmonic oscillator system.

The absorption spectra plotted below all have the same integrated intensity, however their shapes are altered because of the differing extent of displacement of the excited state potential energy surface.

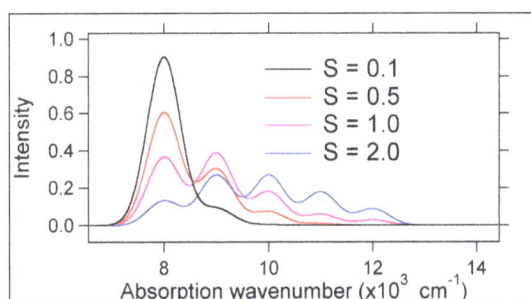

Figure: Stick spectra, dressed with Gaussians, for the small to large displacements in harmonic oscillator system.

So the nature of the relative vibronic band intensities can tell us whether there is a displacement of the equilibrium nuclear coordinate that accompanied a transition. When will there be an increase in bond length (i.e., $Q_e > R_e$)? This occurs when an electron is promoted from a bonding molecular orbital to a non-bonding or anti-bonding molecular orbitals (i.e., when the bond order is less in the excited state than the ground state).

- Non-bonding molecular orbital $\rightarrow\rightarrow$ bonding molecular orbital.

- Anti-bonding molecular orbital $\rightarrow\rightarrow$ bonding molecular orbital.

- Anti-bonding molecular orbital $\rightarrow\rightarrow$ non-bonding molecular orbital.

In short, when the bond order is higher in the excited state than the ground state, then $Q_e > R_e$. A decrease in bondlength will occur when the opposite happens.

Interpretation of Electron Spectroscopy

Often, during electronic spectroscopy, the electron is excited first from an initial low energy state to a higher state by absorbing photon energy from the spectrophotometer. If the wavelength of the incident beam has enough energy to promote an electron to a higher level, then we can detect this in the absorbance spectrum. Once in the excited state, the electron has higher potential energy and will relax back to a lower state by emitting photon energy. This is called fluorescence and can be detected in the spectrum as well.

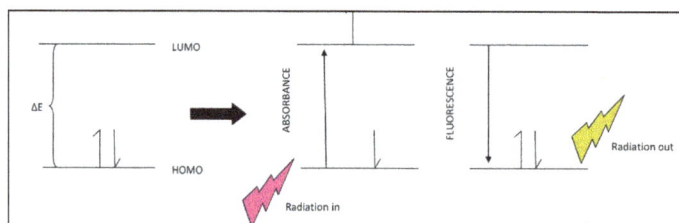

Embedded into the electronic states (n=1,2,3...) are vibrational levels (v=1,2,3...) and within these are rotational energy levels (j=1,2,3...). Often, during electronic transitions, the initial state may have the electron in a level that is excited for both vibration and rotation. In other words, n=0, v does not = 0 and r does not =0. This can be true for the ground state and the excited state. In addition, due to the Frank Condon Factor, which describes the overlap between vibrational states of two electronic states, there may be visible vibrational bands within the absorption bands. Therefore, vibrational fine structure that can be seen in the absorption spectrum gives some indication of the degree of Frank Condon overlap between electronic states.

When interpreting the absorbance and fluorescence spectra of a given molecule, compound, material, or an elemental material, understanding the possible electronic transitions is crucial. Assigning the peaks in the absorption spectrum can become easier when considering which transitions are allowed by symmetry, the Laporte Rules, electron spin, or vibronic coupling. Knowing the degree of allowedness, one can estimate the intensity of the transition, and the extinction coefficient associated with that transition. These guidelines are a few examples of the selection rules employed for interpreting the origin of spectral bands. Only a complete model of molecular energy diagrams for the species under investigation can make clear the possible electronic transitions. Every different compound will have unique energy spacing between electronic levels, and depending on the type of compound, one can categorize these spacings and find some commonality. For example, aromatic compounds pi to pi* and n to pi* transitions where as inorganic compounds can have similar transitions with Metal to Ligand Charge Transfer (MLCT) and Ligand to Metal Charge Transfer (LMCT) in addition to d-d transitions, which lead to the bright colors of transition metal complexes. Although surprises in science often lead to discovery, it is more fortuitous for the interpreter to predict the spectra rather than being baffled by the observation.

This includes an understanding of the molecular or elemental electronic state symmetries, Russell-Sanders states, spin multiplicities, and forbidden and allowed transitions of a given species.

As the light passes through the monochrometer of the spectrophotometer, it hits the sample with some wavelength and corresponding energy. The ratio of the initial intensity of this light and the final intensity after passing through the sample is measured and recorded as absorbance (Abs). When absorbance is measured at different wavelengths, an absorbance spectrum of Abs vs wavelength can be obtained. This spectra reveals the wavelengths of light that are absorbed by the chemical specie, and is specific for each different chemical. Many electronic transitions can be visible in the spectrum if the energy of the incident light matches or surpasses the quantum of energy separating the ground state and that particular excited state. An example of an absorbance spectrum is given below.

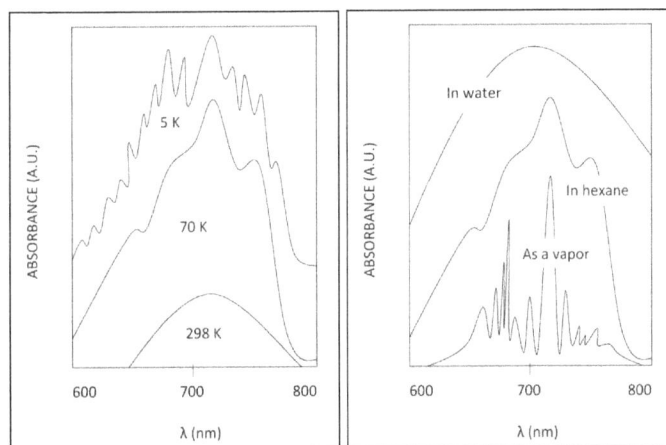

Temperature Effects

Here we can see the effect of temperature and also the effect of solvents on the clarity of the spectrum. We can see from these anonymous compounds that decreasing the temperature allows the vibrational fine structure to emerge. These vibrational bands embedded within the electronic bands represent the transitions from v=n to v'=n. Generally, the v=0 to v'=0 transition is the one with the lowest frequency. From there, increasing energy, the transitions can be from v=0 to v'=n, where n=1,2,3... With a higher temperature, the vibrational transitions become averaged in the spectrum due to the presence of vibrational hot bands and Fermi Resonance, and with this, the vibrational fine structure is lost at higher temperatures.

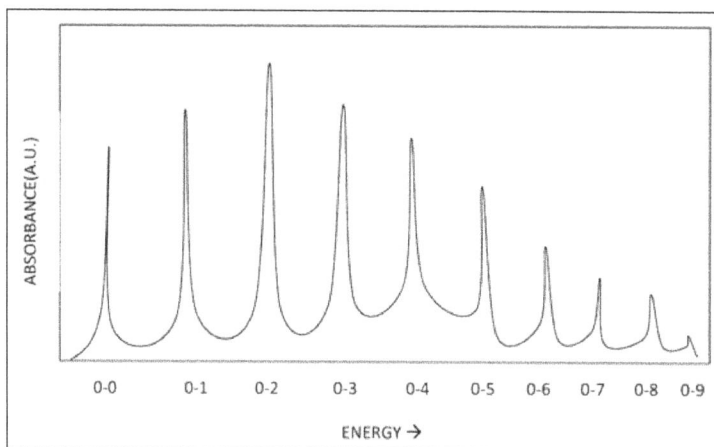

On the left, we see the electronic absorption of an anonymous diatomic molecule. The bands represent the jumping of the electron from the ground electronic state/ground vibrational state n=0/v=0 to the excited electronic state/ground vibrational state n=1/v=0. This is denoted as 0-0 and is defined as the lowest energy transition. The other transitions involve an electronic transition in combination with vibrational transitions within the excited electronic state. In other words, the bands labeled 0-1, 0-2,..., etc. represent the electrons that travel higher in energy than the purely electronic transition (0-0). This introduces the possibility of vibrational information within the electronic spectrum. In fact, due to the Frank Condon overlap of the ground state and excited state, the 0-2 transition is the most probable of all of the transitions. We can see this in the spectra as the 0-2 has the greatest absorbance. If the potential energy wells of the ground and excited states were perfectly eclipsed on top of one another, then the 0-0 transition would be the most probable transition with the highest intensity band. This however, as we have seen, is not often the case. In deed the equilibrium bond length tends to increase as the energy of the state increases usually due to antibonding nature of the state increasing the intramolecular bond length. This causes increased Frank Condon overlap for some transitions and decreased for others as shown in figure above.

Solvent Effects

The effect that the solvent plays on the absorption spectrum is also very important. It is clear that polar solvents give rise to broad bands, non-polar solvents show more resolution, though, completely removing the solvent gives the best resolution. This is due to solvent-solute interaction. The solvent can interact with the solute in its ground state or excited state through intermolecular bonding. For example, a polar solvent like water has the ability of hydrogen bonding with the solute if the solute has a hydrogen bonding component, or simply through induced dipole-dipole interactions. The non-polar solvents can interact though polarizability via London interactions also causing a blurring of the vibronic manifold. This is due to the solvent's tendency to align its dipole moment with the dipole moment of the solute. Depending on the interaction, this can cause the ground state and the excited state of the solute to increase or decrease, thus changing the frequency of the absorbed photon. Due to this, there are many different transition energies that become average together in the spectra. This causes peak-broadening. The effects of peak broadening are most severe for polar solvent, less so for non-polar solvents, and absent when the solute is in vapor phase.

Group Theory and the Transition Moment Integral

When estimating the intensities of the absorption peaks, we use the molar absorptivity constant (epsilon). If the transition is "allowed" then the molar absorptivity constant from the Beer's Law Plot will be high. This means that the probability of transition is large. If the transition is not allowed, then there will be no intensity and no peak on the spectrum. Transitions can be "partially allowed" as well, and these bands appear with a lower intensity than the full allowed transitions. One way to decide whether a transition will be allowed or not is to use symmetry arguments with Group Theory.

If the symmetries of the ground and final state of a transition are correct, then the transition is symmetry allowed. We express this by modifying the transition moment integral from an integral of eigenstates to an orthogonally expressed direct product of the symmetries of the states.

$$\int \psi_2 \mu \psi_1 \, d\tau \rightarrow \Gamma_2 \otimes \Gamma \mu_{xyz} \otimes \Gamma_1$$

$$\int \psi_{v_2} \psi_{el_2} \mu \psi_{el_1} \, d\tau \rightarrow \Gamma_{v_2} \otimes \Gamma_{el_2} \otimes \Gamma \mu_{xyz} \otimes \Gamma_{v_1} \otimes \Gamma_{el_1}$$

The conversions of integration to direct products of symmetry as shown gives spectroscopists a short cut into deciding whether the transition will be allowed or forbidden. A transition will be forbidden if the direct products of the symmetries of the electronic states with the coupling operator is odd. More specifically, if the direct product does not contain the totally symmetric representation, then the transition is forbidden by symmetry arguments. If the product does contain the totally symmetric representation $(A, A_1, A_{1g}...$etc) then the transition is symmetry allowed.

Some transitions are forbidden by the equation $\int \psi_2 \mu \psi_1 \, d\tau \rightarrow \Gamma_2 \otimes \Gamma \mu_{xyz} \otimes \Gamma_1$ and one would not expect to be able to see the band that corresponds to the transition; however, a weak absorbance band is quite clear on the spectrum of many compounds. The transition may be forbidden via pure electronic symmetries; however, for an octahedral complex for example since it has a center of inversion, the transition is weakly allowed because of vibronic coupling. When the octahedra of a transition metal complex is completely symmetric (without vibrations), the transition cannot occur. However, when vibrations exist, they temporarily perturb the symmetry of the complex and allow the transition by equation $\int \psi_{v_2} \psi_{el_2} \mu \psi_{el_1} \, d\tau \rightarrow \Gamma_{v_2} \otimes \Gamma_{el_2} \otimes \Gamma \mu_{xyz} \otimes \Gamma_{v_1} \otimes \Gamma_{el_1}$. If the product of all of these representations contains the totally symmetric representation, then the transition will be allowed via vibronic coupling even if it forbidden electronically.

Hot Bands

Some transitions are forbidden by symmetry and do not appear in the absorption spectrum. If the symmetries are correct, then another state besides the ground state can be used to make the otherwise forbidden transition possible. This is accomplished by hot bands, meaning the electrons in the ground state are heated to a higher energy level that has a different symmetry. When the transition moment integral is solved with the new hot ground state, then the direct product of the symmetries may contain the totally symmetric representation. If we employ the old saying, "You can't get there from here" then we would be referring to the transition from the ground state to the excited state. However, if we thermally excite the molecules from out of the ground state, then, "we can get there from here."

Knowing whether a transition will be allowed by symmetry is an essential component to interpreting the spectrum. If the transition is allowed, then it should be visible with a large extinction coefficient. If it is forbidden, then it should only appear as a weak band if it is allowed by vibronic coupling. In addition to this, a transition can also be spin forbidden. The examples below of excited state symmetries, give an indication of what spin forbidden means:

$^1A_{1g} \longrightarrow {}^3E_{1u}$ (spin forbidden)

$^1A_{1g} \longrightarrow {}^3B_{1u}$ (spin forbidden)

$^1A_{1g} \longrightarrow {}^3B_{2u}$ (spin forbidden)

$^1A_{1g} \longrightarrow {}^1E_{1u}$ (spin allowed)

$^1A_{1g} \longrightarrow {}^1B_{1u}$ (spin allowed)

$^1A_{1g} \longrightarrow {}^1B_{2u}$ (spin allowed)

These states are derived from the electron configuration of benzene. Once we have the molecular orbital energy diagram for benzene, we can assign symmetries to each orbital arrangement of the ground state. From here, we can excite an electron from the Highest Occupied Molecular Orbital (HOMO) to the Lowest Unoccupied Molecular Orbital (LUMO). This is the lowest energy transition. Other transitions include moving the electron above the LUMO to higher energy molecular orbitals. To solve for the identity of the symmetry of the excited state, one can take the direct product of the HOMO symmetry and the excited MO symmetry. This give a letter (A, B, E..) an the subscript (1u, 2u, 1g...). The superscript is the spin multiplicity, and from single electron transitions, the spin multiplicity is $2S+1 = M$, where $S = 1$ with two unpaired electrons having the same spin and $S=0$ when the excited electron flips its spin so that the two electrons have opposite spin. This gives $M=1$ and $M=3$ for benzene above. From the results above, we have three transitions that are spin allowed and three that are spin forbidden.

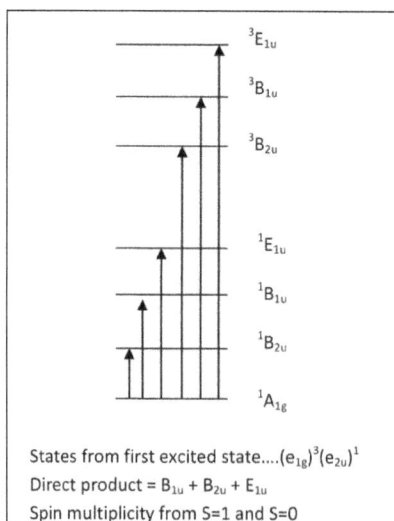

States from first excited state....$(e_{1g})^3(e_{2u})^1$
Direct product = $B_{1u} + B_{2u} + E_{1u}$
Spin multiplicity from S=1 and S=0

Once we take the direct product of the symmetries and the coupling operator for each of these states given above, we find that only the A_{1g} to E_{1u} transition is allowed by symmetry. Therefore, we have information regarding spin and symmetry allowedness and we have an idea of what the spectra will look like.

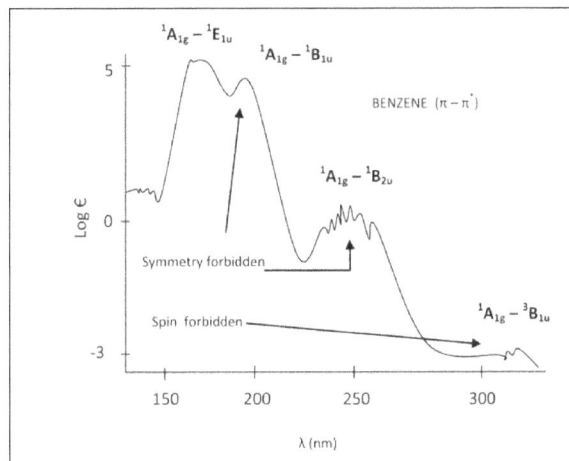

When interpreting the spectrum, it is clear that some transitions are more probable than others. According to the symmetry of excited states, we can now order them from low energy to high energy based on the position of the peaks (E_{1u} is the highest, then B_{1u}, and B2u is lowest). The A_{1g} to E_{1u} transition is fully allowed and therefore the most intense peak. The A_{1g} to B_{1u} and A_{1g} to B_{2u} transitions are symmetry forbidden and thus have a lower probability which is evident from the lowered intensity of their bands. The singlet A_{1g} to triplet B_{1u} transition is both symmetry forbidden and spin forbidden and therefore has the lowest intensity. This transition is forbidden by spin arguments; however, a phenomenon known as spin-orbit coupling can allow this transition to be weakly allowed as well. If spin-orbit coupling exists, then the singlet state has the same total angular momentum as the triplet state so the two states can interact. A small amount of singlet character in the triplet state leads to a transition moment integral that is non-zero, so the transition is allowed.

Organic Molecule Spectra

From the example of benzene, we have investigated the characteristic pi to pi* transitions for aromatic compounds. Now we can move to other organic molecules, which involves n to pi* as well as pi to pi*. Two examples are given below:

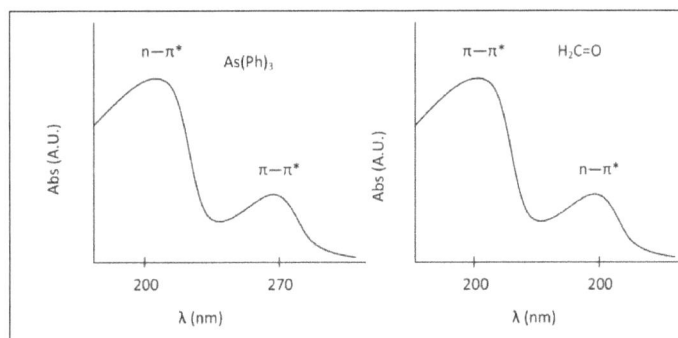

The highest energy transition for both of these molecules has an intensity around 10,000 cm^{-1} and the second band has an intensity of approximately 100 cm^{-1}. In the case of formaldehyde, the n to pi* transition is forbidden by symmetry where as the pi to pi* is allowed. The opposite is true for As(Ph)$_3$ and the difference in molar absorptivity is evidence of this.

n → π Transitions

These transitions involve moving an electron from a nonbonding electron pair to a antibonding π^* orbital. They tend to have molar absorbtivities less than 2000 and undergo a blue shift with solvent interactions (a shift to higher energy and shorter wavelengths). This is because the lone pair interacts with the solvent, especially a polar one, such that the solvent aligns itself with the ground state. When the excited state emerges, the solvent molecules do not have time to rearrange in order to stabilize the excited state. This causes a lowering of energy of the ground state and not the excited state. Because of this, the energy of the transition increases, hence the "blue shift".

π → π Transitions

These transitions involve moving an electron from a bonding π orbital to an antibonding π^* orbital. They tend to have molar absorptivities on the order of 10,000 and undergo a red shift with solvent interactions (a shift to lower energy and longer wavelengths). This could either be due to a raising of the ground state energy or lowering of the excited state energy. If the excited state is polar, then it will be solvent stabilized, thus lowering its energy and the energy of the transition.

Inorganic Molecule Spectra

Speaking of transition probabilities in organic molecules is a good seq way into interpreting the spectra of inorganic molecules. Three types of transitions are important to consider are Metal to Ligand Charge Transfer (MLCT), Ligand to Metal Charge Transfer (LMCT), and d-d transitions. To understand the differences of these transitions we must investigate where these transitions originate. To do this, we must define the difference between pi accepting and pi donating ligands.

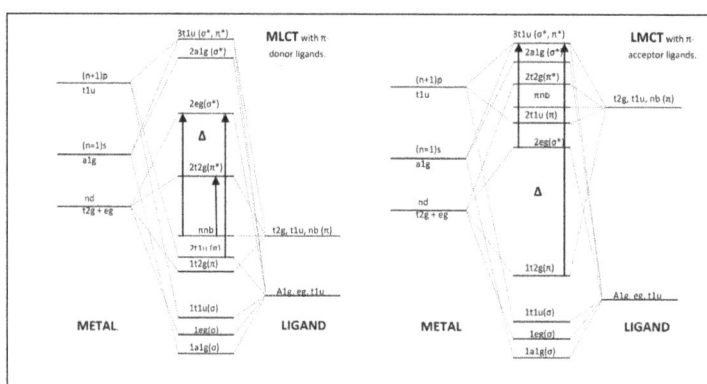

d-d Transitions

From these two molecular orbital energy diagrams for transition metals, we see that the pi donor ligands lie lower in energy than the pi acceptor ligands. According to the spectral chemical series, one can determine whether a ligand will behave as a pi accepting or pi donating. When the ligand is more pi donating, its own orbitals are lower in energy than the t2g metal orbitals forcing the frontier orbitals to involve an antibonding pi* (for t2g) and an antibonding sigma* (for eg). This is in contrast to the pi accepting ligands which involve a bonding pi (t2g) and an antibonding sigma* (eg). Because of this, the d-d transition (denoted above by delta) for the pi acceptor ligand complex is larger than the pi donor ligand. In the spectra, we would see the d-d transitions of pi acceptor ligands to be of a higher frequency than the pi donor ligands. In general though, these transitions appear as weakly intense on the spectrum because they are Laporte forbidden. Due to vibronic coupling; however, they are weakly allowed and because of their relatively low energy of transition, they can emit visible light upon relaxation which is why many transition metal complexes are brightly colored. The molar extinction coefficients for these transition hover around 100.

LMCT Transitions

At an even higher energy are the LMCT which involve pi donor ligands around the metal. These transitions arise because of the low-lying energy of the ligand orbitals. Therefore, we can consider this as a transition from orbitals that are ligand in character to orbitals that are more metal in character, hence the name, Ligand to Metal Charge Transfer. The electron travels from a bonding pi or non-bonding pi orbital into a sigma* orbital. These transitions are very strong and appear very intensely in the absorbance spectrum. The molar extinction coefficients for these transitions are around 10^4. Examples of pi donor ligands are as follows: F^-, Cl^-, Br-, I^-, H_2O, OH^-, RS^-, S_2^-, NCS^-, NCO^-, etc.

MLCT Transitions

The somewhat less common MLCT has the same intensity and energy of the LMCT as they involve the transition of an electron from the t2g (pi) and the eg (sigma*) to the t1u (pi*/sigma*). These transitions arise from pi acceptor ligands and metals that are willing to donate electrons into the orbitals of Ligand character. This is the reason that they are less frequent since metals commonly accept electrons rather than donate them. All the same, both types of Charge Transfer bands are more intense than d-d bands since they are not Laporte Rule forbidden. Examples of pi accepting ligands are as follows: CO, NO, CN^-, N_2, bipy, phen, RNC, $C_5H_5^-$, C=C double bonds, C=C triple bonds.

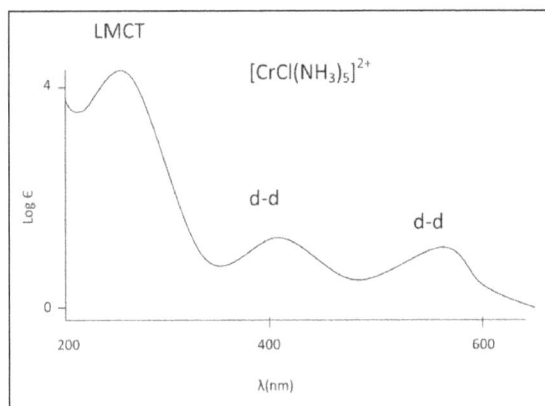

From this spectra of an octahedral Chromium complex, we see that the d-d transitions are far weaker than the LMCT. Since Chlorine is a pi donor ligand in this example, we can label the CT band as LMCT since we know the electron is transitioning from a MO of ligand character to a MO of metal character. The Laporte forbidden (symmetry forbidden) d-d transitions are shown as less intense since they are only allowed via vibronic coupling.

In addition, the d-d transitions are lower in energy than the CT band because of the smaller energy gap between the t2g and eg in octahedral complexes (or eg to t2g in tetrahedral complexes) than the energy gap between the ground and excited states of the charge transfer band.

These transitions abide by the same selection rules that organic molecules follow: spin selection and symmetry arguments. The Tanabe and Sugano diagrams for transition metal complexes can be a guide for determining which transitions are seen in the spectrum. We will use the $[CrCl(NH_3)_5]^{2+}$ ion as an example for determining the types of transitions that are spin allowed. To do this we look up the Tanabe and Sugano diagrams for Octahedral fields. Since Cr in the complex has three electrons, it is a d^3 and so we find the diagram that corresponds to d^3 metals.

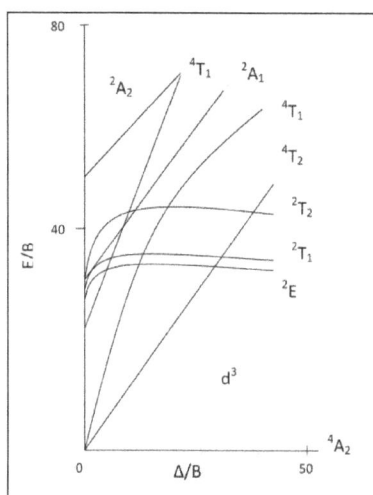

Based on the TS diagram on the left, and the information we have already learned, can you predict which transition will be spin allowed and which ones will be forbidden? From the diagram we see that the ground state is a $^4A_2$. This is because of the three unpaired electrons which make $M=2S+1=4$. The A comes from the fact that there is only one combination of electrons possible. With a spin multiplicity of 4, by the spin selection rules, we can only expect intense transitions between the ground state $^4A_2$ and $^4T_2$, $^4T_1$, and the other $^4T_1$ excited state. The other transitions are spin forbidden. Therefore, we would expect to see three d-d transitions on the absorption spectra.

For us to visualize this, we can draw these transitions in order of increasing energy and then plot the spectrum as we would expect it for only the d-d transitions in a d^3 octahedral complex.

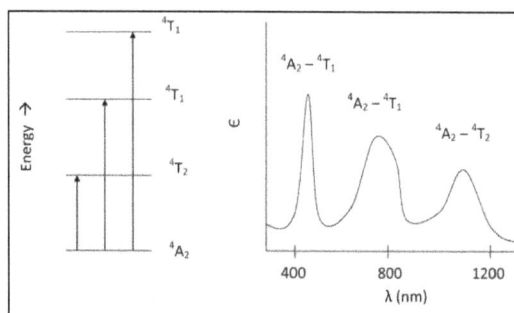

From three spin allowed transitions, we would expect to see three d-d bands appear on the spectrum. In addition to these of course, the LMCT band will appear as well.

Fluorescence

Now that we have discussed the nature of absorption involving an electron absorbing photon energy to be excited to a higher energy level, now we can discuss what happens to that excited electron. Due to its higher potential energy, the electron will relax back to its initial ground state, and in the process, emit electromagnetic radiation. The energy gap between the excited state and the state to which the electron falls determines the wavelength of light that will be emitted. This process is called fluorescence. Generally, the wavelengths of fluorescence are longer than absorbance, can you explain why? Given the following diagram, one can see that vibrational relaxation occurs in the excited electronic state such that the electronic relaxation occurs from the ground vibrational state of the excited electronic state. This causes lower energy electronic relaxations than the previous energy of absorption.

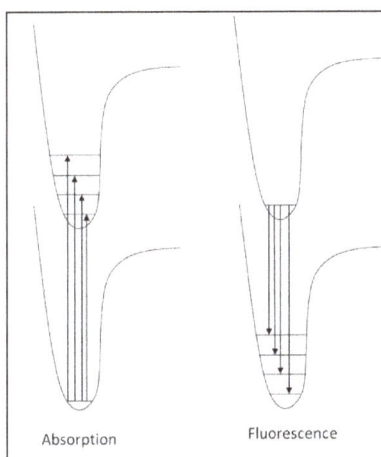

Here we see that the absorption transitions by default involve a greater energy change than the emission transitions. Due to vibrational relaxation in the excited state, the electron tends to relax only from the v'=0 ground state vibrational level. This gives emission transitions of lower energy and consequently, longer wavelength than absorption. When obtaining fluorescence, we have to block out the transmitted light and only focus on the light being emitted from the sample, so the detector is usually 90 degrees from the incident light. Because of this emission spectra are generally obtained separately from the absorption spectra; however, they can be plotted on the same graph as shown.

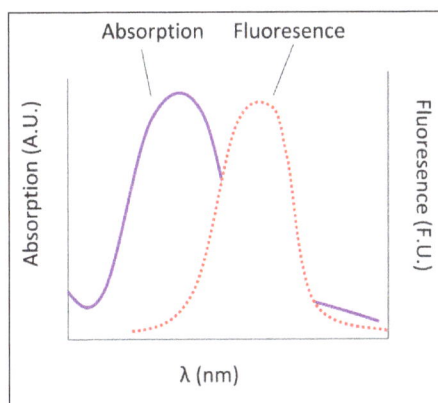

Generally separated by ~10nm, the fluorescence peak follows the absorption peak according to the spectrum.

Fluorescence and Phosphorescence

Fluorescence and phosphorescence are types of molecular luminescence methods. A molecule of analyte absorbs a photon and excites a species. The emission spectrum can provide qualitative and quantitative analysis. The term fluorescence and phosphorescence are usually referred as photoluminescence because both are alike in excitation brought by absorption of a photon. Fluorescence differs from phosphorescence in that the electronic energy transition that is responsible for fluorescence does not change in electron spin, which results in short-live electrons ($< 10^{-5}$ s)in the excited state of fluorescence. In phosphorescence, there is a change in electron spin, which results in a longer lifetime of the excited state (second to minutes). Fluorescence and phosphorescence occurs at longer wavelength than the excitation radiation.

Fluorescence can occur in gaseous, liquid, and solid chemical systems. The simple kind of fluorescence is by dilute atomic vapors. A fluorescence example would be if a 3s electron of a vaporized sodium atom is excited to the 3p state by absorption of a radiation at wavelength 589.6 and 589.0 nm. After 10-8 s, the electron returns to ground state and on its return it emits radiation of the two wavelengths in all directions. This type of fluorescence in which the absorbed radiation is remitted without a change in frequency is known as resonance fluorescence. Resonance fluorescence can also occur in molecular species. Molecular fluorescence band centers at wavelengths longer than resonance lines. The shift toward longer wavelength is referred to as the Stokes Shift.

Excited States Produces Fluorescence and Phosphorescence

Understanding the difference between fluorescence and phosphorescence requires the knowledge of electron spin and the differences between singlet and triplet states. The topics below lay out a general review of the rules and descriptions of electron spin and states.

Electron Spin

The Pauli Exclusion principle states that two electrons in an atom cannot have the same four quantum numbers (n, l, ml, ms) and only two electrons can occupy each orbital where they must have opposite spin states. These opposite spin states are called spin pairing. Because of this spin pairing, most molecules do not exhibit a magnetic field and are diamagnetic. In diamagnetic molecules, electrons are not attracted or repelled by the static electric field. Free radicals are paramagnetic because they contain unpaired electrons have magnetic moments that are attracted to the magnetic field.

Singlet and Triplet Excited State Singlet state is defined when all the electron spins are paired in the molecular electronic state and the electronic energy levels do not split when the molecule is exposed into a magnetic field. A doublet state occurs when there is an unpaired electron that gives two possible orientations when exposed in a magnetic field and imparts different energy to the system. A singlet or a triplet can form when one electron is excited to a higher energy level. In an excited singlet state, the electron is promoted in the same spin orientation as it was in the ground state (paired). In a triplet excited stated, the electron that is promoted has the same spin orientation (parallel) to the other unpaired electron. The difference between the spins of ground singlet, excited singlet, and excited triplet is shown in figure. Singlet, doublet and triplet is derived using the equation for multiplicity, 2S+1, where S is the total spin angular momentum (sum of all the electron spins). Individual spins are denoted as spin up (s = +1/2) or spin down (s = -1/2). If we were to calculated the S for the excited singlet state, the equation would be $2(+1/2 + -1/2)+1 = 2(0)+1 = 1$, therefore making the center orbital in the figure a singlet state. If the spin multiplicity for the excited triplet state was calculated, we obtain $2(+1/2 + +1/2)+1 = 2(1)+1 = 3$, which gives a triplet state as expected.

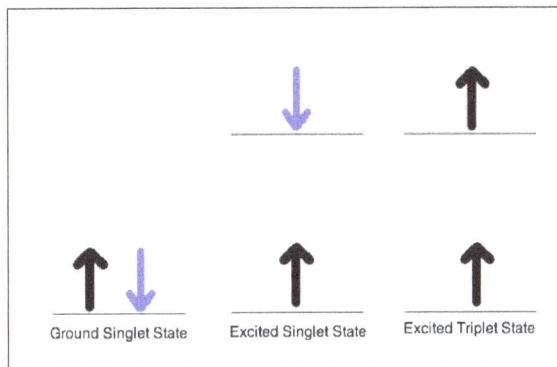

Figure above define Spin in the ground and excited states. The difference between a molecule in the ground and excited state is that the electrons is diamagnetic in the ground state and paramagnetic in the triplet state. This difference in spin state makes the transition from singlet to triplet (or triplet to singlet) more improbable than the singlet-to-singlet transitions. This singlet to triplet (or reverse) transition involves a change electronically excited state. For this reason, the lifetime of the triplet state is longer the singlet state by approximately 104 seconds fold difference. The radiation that induced the transition from ground to excited triplet state has a low probability of occurring, thus their absorption bands are less intense than singlet-singlet state absorption. The excited triplet state can be populated from the excited singlet state of certain molecules which results in phosphorescence. These spin multiplicities in ground and excited states can be used to explain transition in photoluminescence molecules by the Jablonski diagram.

Photoluminescent Energy Level Diagrams

The Jablonski diagram that drawn below is a partial energy diagram that represents the energy of photoluminescent molecule in its different energy states. The lowest and darkest horizontal line represents the ground-state electronic energy of the molecule which is the singlet state labeled as So. At room temperature, majority of the molecules in a solution are in this state.

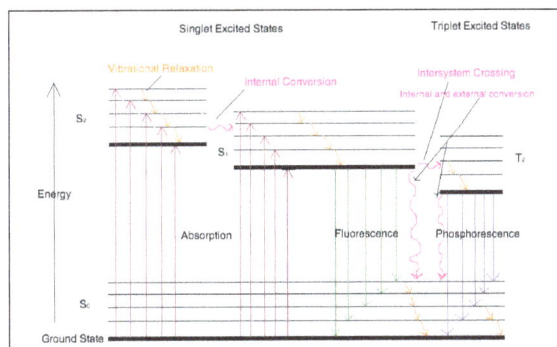

Figure: Partial Jablonski Diagram for Absorption, Fluorescence and Phosphorescence

The upper lines represent the energy state of the three excited electronic states: S_1 and S_2 represent the electronic singlet state (left) and T_1 represents the first electronic triplet state (right). The upper darkest line represents the ground vibrational state of the three excited electronic state. The energy of the triplet state is lower than the energy of the corresponding singlet state. There are numerous vibrational levels that can be associated with each electronic state as denoted by the thinner lines. Absorption transitions (red lines in figure) can occur from the ground singlet electronic state (So) to various vibrational levels in the singlet excited vibrational states. It is unlikely that a transition from the ground singlet electronic state to the triplet electronic state because the electron spin is parallel to the spin in its ground state. This transition leads to a change in multiplicity and thus has a low probability of occurring which is a forbidden transition. Molecules also go through vibration relaxation to lose any excess vibrational energy that remains when excited to the electronic states (S_1 and S_2) as demonstrated in orange in figure. The knowledge of forbidden transition is used to explain and compare the peaks of absorption and emission.

Absorption and Emission Rates

The table below compares the absorption and emission rates of fluorescence and phosphorescence. The rate of photon absorption is very rapid. Fluorescence emission occurs at a slower rate. Since the triplet to singlet (or reverse) is a forbidden transition, meaning it is less likely to occur than the singlet-to-singlet transition, the rate of triplet to singlet is typically slower. Therefore, phosphorescence emission requires more time than fluorescence.

Table: Rates of Absorption and Emission comparison.

Absorption/Emission	Rate (seconds)	Comments
Photon Absorption	10^{-14} to 10^{-15}	Fast
Fluorescence emission	10^{-5} to 10^{-10}	Fast (singlet to singlet transition)
Phosphorescence Emission	10^{-4} to 10	Slow (forbidden transition)

Deactivation Processes

A molecule that is excited can return to the ground state by several combinations of mechanical steps that will be described below and shown in figure. The deactivation process of fluorescence and phosphorescence involve an emission of a photon radiation as shown by the straight arrow in figure. The wiggly arrows in figure are deactivation processes without the use of radiation. The favored deactivation process is the route that is most rapid and spends less time in the excited state. If the rate constant for fluorescence is more favorable in the radiationless path, the fluorescence will be less intense or absent.

Vibrational Relaxation

A molecule maybe to promoted to several vibrational levels during the electronic excitation process. Collision of molecules with the excited species and solvent leads to rapid energy transfer and a slight increase in temperature of the solvent. Vibrational relaxation is so rapid that the lifetime of a vibrational excited molecule ($<10^{-12}$) is less than the lifetime of the electronically excited state. For this reason, fluorescence from a solution always involves the transition of the lowest vibrational level of the excited state. Since the space of the emission lines are so close together, the transition of the vibrational relaxation can terminate in any vibrational level of the ground state.

Internal Conversion

Internal conversion is an intermolecular process of molecule that passes to a lower electronic state without the emission of radiation. It is a crossover of two states with the same multiplicity meaning singlet-to-singlet or triplet-to-triplet states. The internal conversion is more efficient when two electronic energy levels are close enough that two vibrational energy levels can overlap as shown in between S_1 and S_2. Internal conversion can also occur between S_0 and S_1 from a loss of energy by fluorescence from a higher excited state, but it is less probable.

The mechanism of internal conversion from S_1 to S_0 is poorly understood. For some molecules, the vibrational levels of the ground state overlaps with the first excited electronic state, which leads to fast deactivation. These usually occur with aliphatic compounds (compound that do not contain ring structure), which would account for the compound is seldom fluorescing. Deactivation by energy transfer of these molecules occurs so rapidly that the molecule does not have time to fluoresce.

External Conversion

Deactivation of the excited electronic state may also involve the interaction and energy transfer between the excited state and the solvent or solute in a process called external conversion. Low temperature and high viscosity leads to enhanced fluorescence because they reduce the number of collision between molecules, thus slowing down the deactivation process.

Intersystem Crossing

Intersystem crossing is a process where there is a crossover between electronic states of different multiplicity as demonstrated in the singlet state to a triplet state (S_1 to T_1) on figure. The probability of intersystem crossing is enhanced if the vibration levels of the two states overlap. Intersystem crossing is most commonly observed with molecules that contain heavy atom such as iodine or bromine. The spin and orbital interaction increase and the spin become more favorable. Paramagnetic species also enhances intersystem crossing, which consequently decreases fluorescence.

Phosphorescence

Deactivation of the electronic excited state is also involved in phosphorescence. After the molecule transitions through intersystem crossing to the triplet state, further deactivation occurs through internal or external fluorescence or phosphorescence. A triplet-to-singlet transition is more probable than a singlet-to-singlet internal crossing. In phosphorescence, the excited state lifetime is inversely proportional to the probability that the molecule will transition back to the ground state. Since the lifetime of the molecule in the triplet state is large (10-4 to 10 second or more), transition is less probable which suggest that it will persist for some time even after irradiation has stopped. Since the external and internal conversion compete so effectively with phosphorescence, the molecule has to be observed at lower temperature in highly viscous media to protect the triplet state.

Variables that affect Fluorescence

After discussing all the possible deactivation processes, variable that affect the emissions to occur. Molecular structure and its chemical environment influence whether a substance will fluoresce and the intensities of these emissions.

Quantum Yield

The quantum yield or quantum efficiency is used to measure the probability that a molecule will fluoresce or phosphoresce. For fluorescence and phosphorescence is the ratio of the number of molecules that luminescent to the total number of excited molecules. For highly fluoresce molecules, the quantum efficiency approaches to one. Molecules that do not fluoresce have quantum efficiencies that approach to zero.

Fluorescence quantum yield (ϕ) for a compound is determined by the relative rate constants (k) of various deactivation processes by which the lowest excited singlet state is deactivated to the ground state. The deactivation processes including fluorescence (kf), intersystem crossing (ki), internal conversion (kic), predissociation (kpd), dissociation (kd), and external conversion (kec) allows one to qualitatively interpret the structural and environmental factors that influence the intensity of the fluorescence. They are related by the quantum yield equation given below:

$$kfkf + ki + kec + kic + kpd + kd$$

Using this equation as an example to explain fluorescence, a high fluorescence rate (k_f) value and low values of the all the other relative rate constant terms ($k_f + k_i + k_{ec} + k_{ic} + k_{pd} + k_d$) will give a large φ, which suggest that fluorescence is enhanced. The magnitude of k_f, k_d, and k_{pd} depend on the chemical structure, while the rest of the constants k_i, k_{ec}, and k_{ic} are strongly influenced by the environment.

Transition Types in Fluorescence

Fluorescence rarely results from absorption of ultraviolet radiation of wavelength shorter than 250 nm because radiation at this wavelength has sufficient energy to deactivate the electron in the excited state by predissociation or dissociation. The bond of some organic molecules would rupture at 140 kcal/mol, which corresponds to 200-nm of radiation. For this reason, $\sigma \rightarrow \sigma*$ transition in fluorescence are rarely observed. Instead, emissions from the less energetic transition will occur which are either $\pi_* \rightarrow \pi$ or $\pi_* \rightarrow n$ transition.

Molecules that are excited electronically will return to the lowest excited state by rapid vibrational relaxation and internal conversion, which produces no radiation emission. Fluorescence arises from a transition from the lowest vibrational level of the first excited electronic state to one of the vibrational levels in the electronic ground state. In most fluorescent compounds, radiation is produced by a $\pi_* \rightarrow \pi$ or $\pi_* \rightarrow n$ transition depending on which requires the least energy for the transition to occur.

Quantum Efficiency and Transition Types

Fluorescence is most commonly found in compounds in which the lowest energy transition is π→π* (excited singlet state) than n→π* which suggest that the quantum efficiency is greater for π→π∗ transitions. The reason for this is that the molar absorptivity, which measures the probability that a transition will occur, of the π→π∗ transition is 100 to 1000 fold greater than n→π∗ process. The lifetime of π→π∗ (10^{-7} to 10^{-9} s) is shorter than the lifetime of n→π∗ (10^{-5} to 10^{-7}).

Phosphorescent quantum efficiency is the opposite of fluorescence in that it occurs in the n→π∗ excited state which tends to be short lived and less suceptable to deactivation than the π→π∗ triplet state. Intersystem crossing is also more probable for π→π∗ excited state than for the n→π∗ state because the energy difference between the singlet and triplet state is large and spin-orbit coupling is less likely to occur.

Fluorescence and Structure

The most intense fluorescence is found in compounds containing aromatic group with low-energy π→π∗ transitions. A few aliphatic, alicyclic carbonyl, and highly conjugated double-bond structures also exhibit fluorescence as well. Most unsubstituted aromatic hydrocarbons fluoresce in solution too. The quantum efficiency increases as the number of rings and the degree of condensation increases. Simple heterocycles such as the structures listed below do not exhibit fluorescence.

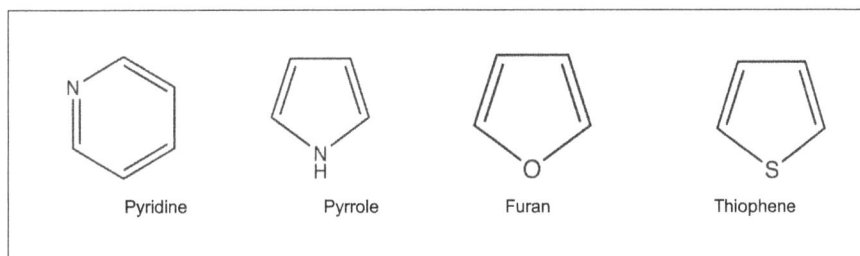

| Pyridine | Pyrrole | Furan | Thiophene |

With nitrogen heterocyclics, the lowest energy transitions is involved in n→π∗ system that rapidly converts to the triplet state and prevents fluorescence. Although simple heterocyclics do not fluoresce, fused-ring structures do. For instance, a fusion of a benzene ring to a hetercyclic structure results in an increase in molar absorptivity of the absorption band. The lifetime of the excited state in fused structure and fluorescence is observed. Examples of fluorescent compounds is shown below.

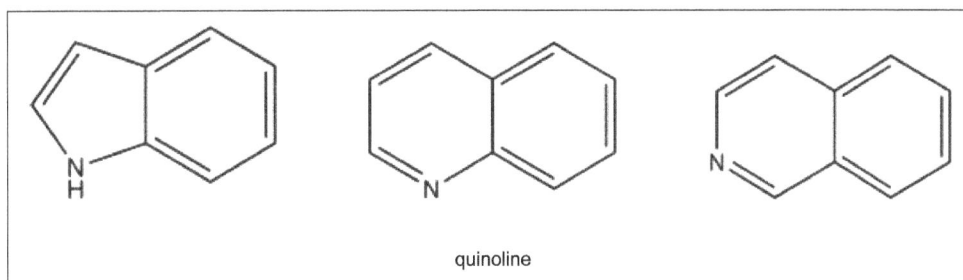

quinoline

Benzene ring substitution causes a shift in the absorption maxima of the wavelength and changes in fluorescence emission. The table below is used to demonstrate and visually show that as benzene is substituted with increasing methyl addition, the relative intensity of fluorescence increases.

Table: Relative intensity of fluorescence comparison with alkane substituted benzenes.

Compound	Structure	Wavelength of Fluorescence (nm)	Relative intensity of Fluorescence
Benzene		270-310	10
Toluene		270-320	17
Propyl Benzene		270-320	17

The relative intensity of fluorescence increases as oxygenated species increases in substitution. The values for such increase is demonstrated in the table below.

Table: Relative intensity of fluorescence comparison with benzene with oxygenated substituted benzene.

Compound	Structure	Wavelength of Fluorescence (nm)	Relative intensity of Fluorescence
Phenol		285-365	18
Phenolate ion		310-400	10
Anisole		285-345	20

Influence of a halogen substitution decreases fluorescence as the molar mass of the halogen increases. This is an example of the "heavy atom effect" which suggest that the probability of intersystem crossing increases as the size of the molecule increases. As demonstrated in the table below, as the molar mass of the substituted compound increases, the relative intensity of the fluorescence decreases.

Table: Relative intensity fluorescence comparison with halogen substituted compounds.

Compound	Structure	Wavelength of Fluorescence (nm)	Relative intensity of Fluorescence
Fluorobenzene		270-320	10
Chlorobenzene		275-345	7

Bromobenzene		290-380	5

In heavy atom substitution such as nitro derivatives or heavy halogen substitution such as iodobenzene, the compounds are subject to predissociation. These compounds have bonds that easily rupture that can then absorb excitation energy and go through internal conversion. Therefore, the relative intensity of fluorescence and fluorescent wavelength is not observed and this is demonstrated in the table below.

Table: Relative fluorescent intensities of iodobenzene and nitro derivative compounds.

Compound	Structure	Wavelength of Fluorescence (nm)	Relative intensity of Fluorescence
Iodobenzene		None	0
Anilinium ion		None	0
Nitrobenzene		None	0

Carboxylic acid or carbonyl group on aromatic ring generally inhibits fluorescence since the energy of the $n \rightarrow \pi^*$ transition is less than $\pi \rightarrow \pi^*$ transition. Therefore, the fluorescence yield from $n \rightarrow \pi^*$ transition is low.

Table: Relative fluorescent intensity of benzoic acid.

Compound	Structure	Wavelength of Fluorescence (nm)	Relative intensity of Fluorescence
Benzoic Acid		310-390	3

Effect of Structural Rigidity on Fluorescence

Fluorescence is particularly favored in molecules with rigid structures. The table below compares the quantum efficiencies of fluorine and biphenyl which are both similar in structure that there is a bond between the two benzene group. The difference is that fluorene is more rigid from the addition methylene bridging group. By looking at the table below, rigid fluorene has a higher quantum efficiency than unrigid biphenyl which indicates that fluorescence is favored in rigid molecules.

Table: Quantum Efficiencies in Rigid vs. Nonrigid structures.

Compound	Structure	Quantum Efficiency
Fluorene	Fluorene	1.0
Biphenyl	biphenyl	1.2

This concept of rigidity was used to explain the increase in fluorescence of organic chelating agent when the compound is complexed with a metal ion. The fluorescence intensity of 8-hydroxyquinoline is much less than its zinc complex.

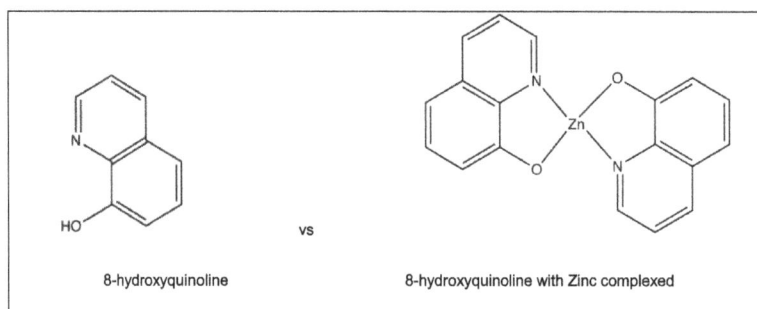

8-hydroxyquinoline   vs   8-hydroxyquinoline with Zinc complexed

The explanation for lower quantum efficiency or lack of rigidity in caused by the enhanced internal conversion rate (k_{ic}) which increases the probability that there will be radiationless deactivation. Nonrigid molecules can also undergo low-frequency vibration which accounts for small energy loss.

Temperature and Solvent Effects

Quantum efficiency of Fluorescence decreases with increasing temperature. As the temperature increases, the frequency of the collision increases which increases the probability of deactivation by external conversion. Solvents with lower viscosity have higher possibility of deactivation by external conversion. Fluorescence of a molecule decreases when its solvent contains heavy atoms such as carbon tetrabromide and ethyl iodide, or when heavy atoms are substituted into the fluorescing compound. Orbital spin interaction result from an increase in the rate of triplet formation, which decreases the possibility of fluorescence. Heavy atoms are usually incorporated into solvent to enhance phosphorescence.

Effect of pH on Fluorescence

The fluorescence of aromatic compound with basic or acid substituent rings are usually pH dependent. The wavelength and emission intensity is different for protonated and unprotonated forms of the compound as illustrated in the table below.

Table: Quantum efficiency comparison due to protonation.

Compound	Structure	Wavelength of Fluorescence (nm)	Relative intensity of Fluorescence
Aniline		310-405	20
Anilinium ion		None	0

The emission changes of this compound arises from different number of resonance structures associated with the acidic and basic forms of the molecule. The additional resonance forms provides a more stable first excited state, thus leading to fluorescence in the ultraviolet region. The resonance structures of basic aniline and acidic anilinium ion is shown below:

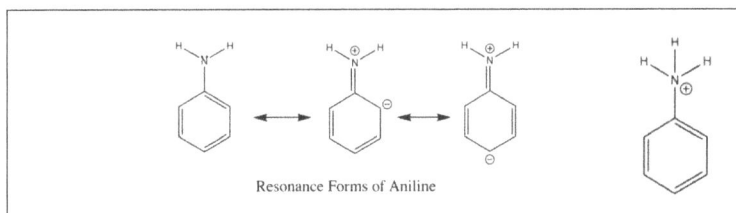

Resonance Forms of Aniline

Basic aniline fluorescence of certain compounds have been used a detection of end points in acid-base titrations. An example of this type of fluorescence seen in compound as a function of pH is the phenolic form of 1-naphthol-4- sulfonic acid. This compound is not detectable with the eye because it occurs in the ultraviolet region, but with an addition of a base, it becomes converted to a phenolate ion, the emission band shifts to the visible wavelength where it can be visually seen. Acid dissociation constant for excited molecules differs for the same species in the ground state. These changes in acid or base dissociation constant differ in four or five orders of magnitude.

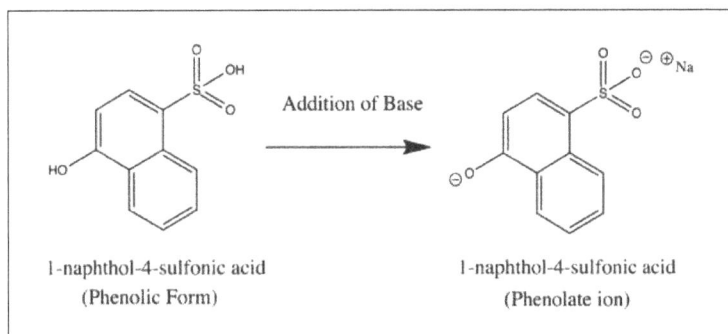

Addition of Base

1-naphthol-4-sulfonic acid
(Phenolic Form)

1-naphthol-4-sulfonic acid
(Phenolate ion)

Dissolved oxygen reduces the intensity of fluorescence in solution, which results from a photochemically induced oxidation of fluorescing species. Quenching takes place from the paramagnetic properties of molecular oxygen that promotes intersystem crossing and conversion of excited molecules to triplet state. Paramagnetic properties tend to quench fluorescence.

Effects of Concentration on Fluorescence Intensity

The power of fluorescence emission F is proportional to the radiant power is proportional to the radiant power of the excitation beam that is absorbed by the system. The equation below best describes this relationship.

$$F = \phi_f K''(P_0 - P) = K'(P_0 - P)$$

Since ϕ f K" is constant in the system, it is represented at K'. The table below defines the variables in this equation.

Table: Definitions of all the variables defined in the Fluorescence Emission (F) in equation $F = \phi_f K''(P_0 - P) = K'(P_0 - P)$.

Variable	Definition
F	Power of fluorescence emission
P_0	Power of incident beam on solution
P	Power after transversing length b in medium
K"	Constant dependent on geometry and other factors
f	Quantum efficiency

Fluorescence emission (F) can be related to concentration (c) using Beer's Law stating:

$$F = \varepsilon bc$$

Where ε is the molar absorptivity of the molecule that is fluorescing. Rewriting equation $F = \varepsilon bc$ gives:

$$P = P_0 \times 10^{-\varepsilon bc}$$

Plugging this equation $P = P_0 \times 10 - \varepsilon bc$ into Equation $F = \phi_f K''(P_0 - P) = K'(P_0 - P)$ and factoring out P_0 gives us this equation:

$$F = K'P_0(1 - 10^{-\varepsilon bc}).$$

The MacLaurin series could be used to solved the exponential term,

$$F = K'P_0[2.303\varepsilon bc - \frac{(2.303\varepsilon bc)^2}{2!} + \frac{(2.303\varepsilon bc)^3}{3!} + \frac{(2.303\varepsilon bc)^4}{4!} + \dots \frac{(2.303\varepsilon bc)^n}{n!}]$$

Given that 2.303 εbc = Absorbance 0.05, all the subsequent terms after the first can be dropped since the maximum error is 0.13%. Using only the first term, Equation 5 can be rewritten as:

$$F = K'P_0 2.303\varepsilon bc$$

Equation $F = K'P_0 2.303\varepsilon bc$ can be expanded to the equation below and simplified to compare the fluorescence emission F with concentration. If the equation below were to be plotted with F versus c, a linear relation would be observed.

$$F = \phi_f K'' P_0 2.303\varepsilon bc$$

If c becomes so great that the Absorbance > 0.05, the higher terms start to become taken into account and the linearity is lost. F then lies below the extrapolation of the straight-line plot. This excessive absorption is the primary absorption. Another cause of this negative downfall of linearity is the secondary absorption when the wavelength of emission overlaps the absorption band. This occurs when the emission transverse the solution and gets reabsorbed by other molecules by analyte or other species in the solution, which leads to a decrease in fluorescence.

Quenching Methods

Dynamic Quenching is a nonradiative energy transfer between the excited and the quenching agent species (Q). The requirements for a successful dynamic quenching are that the two collision species the concentration must be high so that there is a higher possibility of collision between the two species. Temperature and quenching agent viscosity play a role on the rate of dynamic quenching. Dynamic quenching reduces fluorescence quantum yield and the fluorescence lifetime.

Dissolved oxygen in a solution increases the intensity of the fluorescence by photochemically inducing oxidation of the fluorescing species. Quenching results from the paramagnetic properties of molecular oxygen that promotes intersystem crossing and converts the excited molecules to triplet state. Paramagnetic species and dissolved oxygen tend to quench fluorescence and quench the triplet state.

Static quenching occurs when the quencher and ground state fluorophore forms a dark complex. Fluorescence is usually observed from unbound fluorophore. Static quenching can be differentiated from dynamic quenching in that the lifetime is not affected in static quenching. In long range (Förster) quenching, energy transfer occurs without collision between molecules, but dipole-dipole coupling occurs between excited fluorophore and quencher.

Emission and Excitation Spectra

One of the ways to visually distinguish the difference between each photoluminescence is to compare the relative intensities of emission/excitation at each wavelength. An example of the three types of photoluminescence (absorption, fluorescence and phosphorescence) is shown for phenanthrene in the spectrum below. In the spectrum, the luminescent intensity is measure in a wavelength is fixed while the excitation wavelength is varied. The spectrum in red represents the excitation spectrum, which is identical to the absorption spectrum because in order for fluorescence emission to occur, radiation needs to be absorbed to create an excited state. The spectrum in blue represent fluorescence and green spectrum represents the phosphorescence.

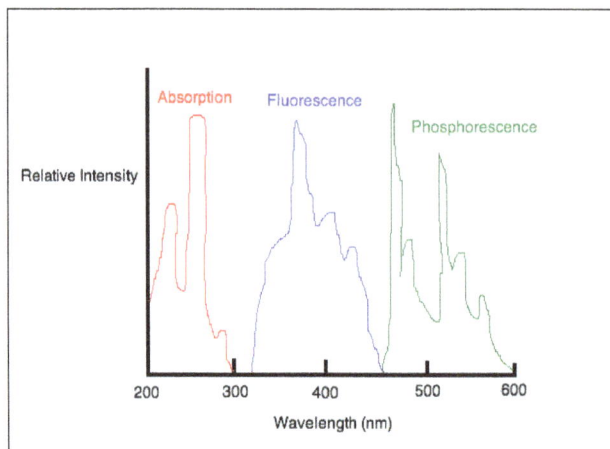

Figure: Wavelength Intensities of Absorption, Fluorescence, and Phosphorescence.

Fluorescence and Phosphorescence occur at wavelengths that are longer than their absorption wavelengths. Phosphorescence bands are found at a longer wavelength than fluorescence band because the excited triplet state is lower in energy than the singlet state. The difference in wavelength could also be used to measure the energy difference between the singlet and triplet state of the molecule. The wavelength (λ) of a molecule is inversely related to the energy (E) by the equation below:

$$E = \frac{hc}{\lambda}$$

As the wavelength increases, the energy of the molecule decrease and vice versa.

Electronic Spectrum

The electronic spectrum covers the range from 200 to 800 nm of the electromagnetic spectrum: 200–400 nm is the ultraviolet region and 400–800 nm is the visible region. Hence, the electronic spectrum is also known as the UV-Vis spectrum. The major application of this spectrum to metal complexes is the determination of the geometry of the complex.

Molecules can also undergo changes in electronic transitions during microwave and infrared absorptions. The energy level differences are usually high enough that it falls into the visible to UV range; in fact, most emissions in this range can be attributed to electronic transitions.

Electron Transitions are not Purely Electronic

We have thus far studied rovibrational transitions--that is, transitions involving both the vibrational and rotational states. Similarly, electronic transitions tend to accompany both rotational and vibrational transitions. These are often portrayed as an electronic potential energy cure with the vibrational level drawn on each curve. Additionally, each vibrational level has a set of rotational levels associated with it.

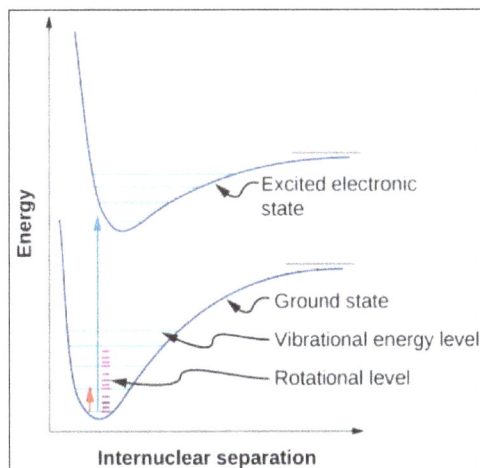

Figure: Three types of energy levels in a diatomic molecule: electronic, vibrational, and rotational. If the vibrational quantum number (n) changes by one unit, then the rotational quantum number (l) changes by one unit.

Recall that in the Born-Oppenheimer approximation, nuclear kinetic energies can be ignored (e.g., fixed) to solve for electronic wavefunctions and energies, which are much faster than rotation or vibration. As such, it is important to note that unlike rovibrational transitions, electronic transitions aren't dependent on rotational or transitional terms and are assumed to be separate. Therefore, when using an anharmonic oscillator-nonrigid rotator approximation (and excluding translation energy), the total energy of a diatomic is:

$$\tilde{E}_{total} = \tilde{v}_{el} + G(v) + F(J)$$

where $\tilde{v}_{el}$ is the electronic transition energy change in wavenumbers, G(n) is the vibrational energy with energy level v (assuming anharmonic oscillator), and F(J) is the rotational energy, assuming a nonrigid rotor. Equation $\tilde{E}_{total} = \tilde{v}_{el} + G(v) + F(J)$ can be expanded accordingly:

$$\underbrace{\tilde{v}_{el}}_{electronic} + \underbrace{\tilde{v}_e\left(v + \frac{1}{2}\right) - \tilde{X}_e\tilde{v}_e\left(v + \frac{1}{2}\right)^2}_{vibrational} + \underbrace{\tilde{B}J(J+1) - \tilde{D}J^2(J+1)^2}_{rotational}$$

Notice that both the vibration constant ($\tilde{v}_e$) and anharmonic constant ($\tilde{X}_e$) are electronic state dependent (and hence the rotational constants would be too, but are ignored here). Since rotational energies tend to be so small compared to electronic, their effects are minimal and are typically ignored when we do calculations and are referred to as vibronic transitions.

The eigenstate-to-eigenstate transitions (e.g., $1 \to 2$) possible are numerous and have absorption lines at,

$$\tilde{v}_{obs} = \tilde{E}_2 - \tilde{E}_1$$

and for simplification, we refer to constants associated with these states as $|'\rangle$ and $|''\rangle$, respectively. So Equation $\tilde{v}_{obs} = \tilde{E}_2 - \tilde{E}_1$ is,

$$\tilde{v}_{obs} = E''\tilde{(}v'') - E'\tilde{(}v')$$

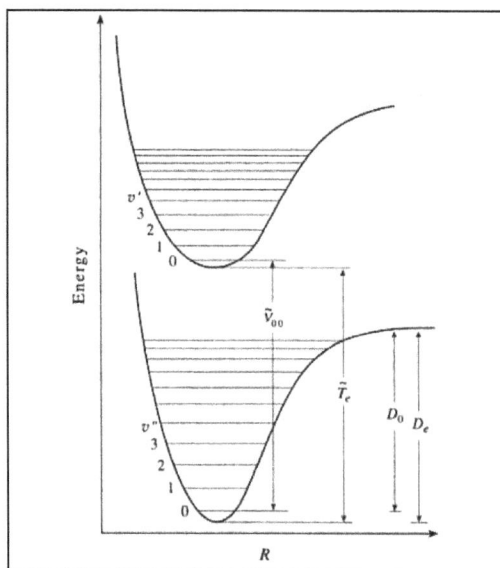

Figure: As you can see, electronic transitions involve two energy potential wells.
The vibrational levels are included, but rotational levels are excluded.

Also important to note that typically vibronic transitions are usually the result of the vibrational $v' = 0$ vibratonal state. Within this assumption and excluding the rotational contributions (due to their low energies), Equation $\underbrace{\tilde{v}_{el}}_{\text{electronic}} \underbrace{+ \tilde{v}_e\left(v + \frac{1}{2}\right) - X_e\tilde{v}_e\left(v + \frac{1}{2}\right)^2}_{\text{vibrational}} + \underbrace{\tilde{B}J(J+1) - \tilde{D}J^2(J+1)^2}_{\text{rotational}}$ can be used with Equation $\tilde{v}_{obs} = \tilde{E}_2 - \tilde{E}_1$ to get:

$$\tilde{v}_{obs} = \tilde{T}_{el} + \left(\frac{1}{2}\tilde{v}'_e - \frac{1}{4}X_e'\tilde{v}'_e\right) - \left(\frac{1}{2}\tilde{v}''_e - \frac{1}{4}X_e''\tilde{v}''_e\right) + \tilde{v}'_e v'' - X_e'\tilde{v}'_e v''(v''+1)$$

A common transition of importance is the $\tilde{v}_{00}$, which is the 0→0 transition and include no vibrational change. For this case, equation:

$$\tilde{v}_{obs} = \tilde{T}_{el} + \left(\frac{1}{2}\tilde{v}'_e - \frac{1}{4}X_e'\tilde{v}'_e\right) - \left(\frac{1}{2}\tilde{v}''_e - \frac{1}{4}X_e''\tilde{v}''_e\right) + \tilde{v}'_e v'' - X_e'\tilde{v}'_e v''(v''+1) \text{ is then,}$$

$$\tilde{v}_{00} = \tilde{T}_{el} + \left(\frac{1}{2}\tilde{v}'_e - \frac{1}{4}X_e'\tilde{v}'_e\right) - \left(\frac{1}{2}\tilde{v}''_e - \frac{1}{4}X_e''\tilde{v}''_e\right)$$

This is the lowest energy possible to observe in an electronic transition although it may be of low intensity.

References

- Electron-spectroscopy-introduction: chemicool.com, Retrieved 2 March, 2019

- Photoelectron-spectroscopy, chemical-engineering: idc-online.com, Retrieved 22 July, 2019

- Centersandinstitutes, engineering-and-computing: sc.edu, Retrieved 4 April, 2019

- Sakurai, J. (1995). Modern Quantum Mechanics (Rev. ed.). Addison-Wesley Publishing Company. p. 332. ISBN 0-201-53929-2

- Briggs, David; Martin P. Seah (1983). Practical Surface Analysis by Auger and X-ray Photoelectron Spectroscopy. Chichester: John Wiley & Sons. ISBN 0-471-26279-X

- Auger-Electron-Spectroscopy, Elemental-Analysis, Physical-Methods-in-Chemistry-and-Nano-Science, Analytical-Chemistry: libretexts.org, Retrieved 14 May, 2019

- Ahn C C (ed.) (2004) Transmission electron energy loss spectrometry in material science and the EELS Atlas, Wiley, Weinheim, Germany, doi:10.1002/3527605495, ISBN 3527405658

- The-Franck-Condon-Principle-Predicts-the-Relative-Intensities-of-Vibronic-Transitions, Physical-and-Theoretical-Chemistry: libretexts.org, Retrieved 17 January, 2019

- Electronic-Spectroscopy, Physical-and-Theoretical-Chemistry: libretexts.org, Retrieved 7 June, 2019

- Electronic-spectroscopy, chemical-engineering: idc-online.com, Retrieved 18 August, 2019

- Electronic-spectra, physics-and-astronomy: sciencedirect.com, Retrieved 9 February, 2019

Laser Spectroscopy

Laser spectroscopy is used for the selective excitation of atomic or molecular species. Some of the techniques used within this field are laser Raman spectroscopy, IR laser spectroscopy, ultrafast laser spectroscopy and laser absorption spectroscopy. This chapter discusses in detail these techniques related to laser spectroscopy.

A branch of spectroscopy in which a laser is used as an intense, monochromatic light source; in particular, it includes saturation spectroscopy, as well as the application of laser sources to Raman spectroscopy and other techniques. Laser spectroscopy is a valuable tool for sensing and chemical analysis.

By employing monochromatic laserradiation, quantum transitions can be induced between specific atomic and molecular energy levels. By contrast, spectroscopy based on nonlaser light sources studies spectra resulting from transitions between an enormous number of quantum states of atoms and molecules.

The first serious laser experiments in spectroscopy were carried out after sufficiently powerful lasers were developed with an output radiation of a fixed frequency in the visible region. Such lasers were used to produce Raman spectra. The advent of the frequency tunable laser opened up fundamentally new possibilities for laser spectroscopy. Through laser spectroscopy it became possible to solve or to attempt to solve important problems that could not be handled by spectroscopy based onconventional light sources.

The high monochromaticity of the radiation from tunable lasers permits the measurement of the true shape of the spectral lines of a substance that is, the shape undistorted by the spread function of the spectroscopic device. This development is particularly important for the spectroscopy of gases in the infrared region, where the resolution of the best industrial devices of the usual type is 0.1 cm, which is 100 times greater than the width of narrow spectral lines.

The methods of nonlinear laser spectroscopy are based on the time and spatial coherence of laser radiation. This coherence makes possible the study of the spectral line structure usually concealed by the Doppler broadening due to the thermalmotion of the gas particles.

Owing to its high monochromaticity and coherence, laser radiation causes a substantial number of particles to move from the ground state to an excited state. As a result, the sensitivity of registration of atoms and molecules is enhanced: Impurities consisting of 10^2 atoms or 10^{10} molecules per cm^3 of a substance can be registered. Methods are being developed forregistering individual atoms and molecules.

Extremely short laser pulses are employed to investigate the rapidly occurring ($\sim 10^{-6}$–10^{-12} sec) processes of excitation, deexcitation and transfer of excitation in a substance. Through the use of pulses of directional laser radiation, the scattering and fluorescence spectra of atoms and molecules in the atmosphere can be studied at substantial distances (-100 km), information can be obtained on the composition of the atmosphere, and environmental pollution can be monitored.

By focusing laser radiation, the composition of small amounts of a substance (with dimensions of the order of a wavelength) can be studied. This technique has been successfully applied to the analysis of local emission spectra.

The devices employed in laser spectroscopy are fundamentally different from conventional spectroscopic devices. In devices using tunable lasers there is no need to separate the radiation into a spectrum by means of dispersing elements (prisms ordiffraction gratings), which are an important part of conventional spectroscopic devices. Sometimes laser spectroscopy makes use of devices in which the radiation is separated into a spectrum by means of nonlinear crystals.

In laser spectroscopy, chemists train a laser beam on a sample, yielding a characteristic light source that can be analyzed by a spectrometer. But laser spectroscopy falls into several different schools, depending on what kind of laser chemists favor and which aspect of an atom's excited response they study. Let's look at some of these more closely.

Named after the scientist who discovered it, C.V. Raman, Raman spectroscopy measures the scattering of monochromatic light caused by a sample. The beam from an argonion laser is directed by a system of mirrors to a lens, which focuses monochromatic light onto the sample. Most of the light bouncing off the sample scatters at the same wavelength as the incoming light, but some of the light does scatter at different wavelengths. This happens because the laser light interacts with phonons, or naturally occurring vibrations present in the molecules of most solid and liquid samples. These vibrations cause the photons of the laser beam to gain or lose energy. The shift in energy gives information about the phonon modes in the system and ultimately about the molecules present in the sample.

Fluorescence refers to the visible radiation emitted by certain substances because of incident radiation at a shorter wave length. In laser-induced fluorescence (LIF), a chemist activates a sample usually with a nitrogen laser alone or a nitrogen laser in combination with a dye laser. The sample's electrons become excited and jump up to higher energy levels. This excitation lasts for a few nanoseconds before the electrons return to their ground state. As they lose energy, the electrons emit light, or fluoresce, at a wavelength longer than the laser wavelength. Because the energy states are unique for each atom and molecule, the fluorescence emissions are discrete and can be used for identification.

LIF is a widely used analytical tool with many applications. For instance, some countries have adopted LIF to protect consumers from pesticide-tainted vegetables. The tool itself consists of a nitrogen laser, a sensor head and a spectrometer, all packaged in a small, portable system. An agricultural inspector directs the laser on a vegetable lettuce leaves, let's say and then analyzes the resulting fluorescence. In some cases, the pesticides can be identified directly. In other cases, they must be identified based on how they interact with chlorophyll, the green pigment present in all leaves.

Laser ablation inductively coupled plasma optical emission spectroscopy (LA-ICP-OES) has a ridiculously complicated name, so let's start with ICP, which is the heart of the analytical technique. The "P" in ICP stands for plasma, an ionized gas consisting of positive ions and free electrons. In nature, plasmas usually form only in stars, where the temperatures are high enough to ionize the gas. But scientists can create plasmas in the lab using something known as a plasma torch. The torch consists of three concentric tubes of silica surrounded by a metal coil. When an electric current passes through the coil, a magnetic field is created, which in turn induces electric currents in a gas, usually argon, allowed to pass through the silica tubes. This excites the argon gas and creates the plasma. A nozzle at the end of the torch acts as an exit for the plasma.

Now the instrument is ready to analyze a sample. In the laser-based version of ICP-OES, a neodymium-doped yttrium aluminum garnet (Nd:YAG) laser is used to cut, or ablate, a few microscopic particles from the sample's surface. That means analysis isn't limited to liquids solids are fair game, as well. The ablated particles are then carried to the plasma torch, where they become excited and emit light.

Laser-induced breakdown spectroscopy (LIBS) is similar to LA-ICP-OES, except that the laser both ablates the sample and creates the plasma. Because LIBS has become increasingly popular in recent years, we're going to give it more attention next.

Laser Raman Spectroscopy

Laser Raman spectroscopy uses a monochromatic laser to interact with molecular vibrational modes in a sample, shifting the laser energy down (Stokes) or up (anti-Stokes) through in elastic scattering. It is most useful for fingerprinting chemical species of which minerals and vein forming fluids are of immediate relevance to all geological investigations and researches. Although Laser Raman spectroscopy (LRS) is most sensitive to covalent bonds with little or no natural dipole moment, as in carbon compounds, it also effectively identifies partially covalent bonds with dipole moments which characterize most known minerals. However, the system, by its basic principle, is ineffective for chemical compounds or minerals that have purely electrovalent or metallic structure.

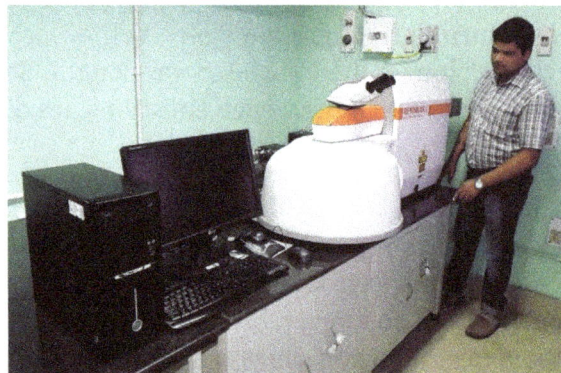

Renishaw in Via Reflex Laser Raman Microscope.

A Renishaw in Via Reflex Laser Raman Microscopies operative at the Central Petrological Laboratory, Geological Survey of India. It is a high-sensitivity system with integrated research grade microscope, enabling high-resolution confocal measurements and supports multiple lasers (540 nm and 785 nm), with automatic software switching of excitation wavelength. The other components of the spectroscope include holographic notch filter (to eliminate the elastic or Rayleigh scatter), lens systems, gratings for resolution and a chargecoupled device CCD. The Raman spectrum collected consists of a sequence of peaks in a wave number vs intensity diagram. Each peak is associated with a different bond in the molecule. This pattern, unique to each substance, is searched against an established library database and the best correlated sample/mineral/fluid is considered the most likely match.

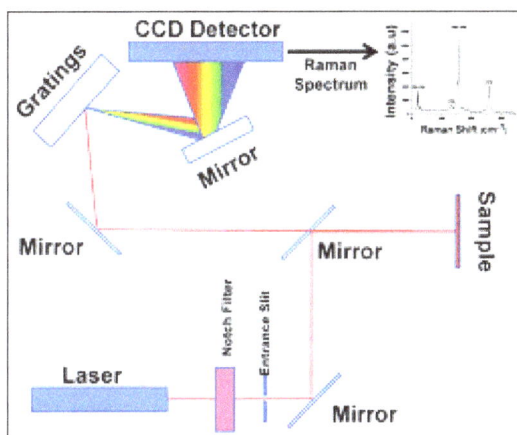

Internal principle of LRS microscope.

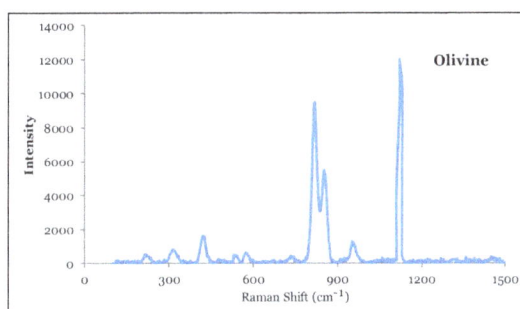

Raman Spectroscopic profile for olivine.

Sample Preparation for LRS

- Laser Raman Spectroscope can easily and effectively operate on a flat surface of a mineral. Such a flat surface includes a cleavage plane or crystal face of a single crystal; a wafer, a cut and polished surface as in gemstone, a polished block of rock (sufficiently small so as to fit on the stage of the microscope) or a thin section.

- A thin polished section or a polished section can give excellent results.

- The sample should not be covered (as with a cover slip) or coated (like gold or carbon coating as done in SEM or EPMA).

- If the sample has already been subjected to study under SEM or EPMA and there exists a layer of coating on the sample, then for best results the coating has to be removed not simply by mechanical rubbing with organic solvent but rather more effectively by repolishing.

- Also carbonate samples should not be Alizarin treated as the dye has its own strong LRS signature.

Procedure for LRS Booking

- GSI officers from Regional Laboratories or from all State Units, who are interested in using the LRS, are requested to send a letter (through proper channel) to the Director, Central Petrological Laboratories, CHQ requesting a time slot.

- The numbers of samples and the type of samples needs to be mentioned in the letter.

- Also required is the e-mail and Fax contacts of the officer concerned.

- The allotted time slot will be sent to the officer via both e-mail and Fax.

- Also it is always preferred that the officer be personally present during the study of his/her samples.

- The knowledge of the officer on his/her specific requirements, textural controls etc. help in both speed and accuracy of the analyses and also helps in minimizing office expenditure.

Applications of LRS in Geological Investigations

Identification of Minerals

Raman Spectroscope is an easy, cost effective and non destructive application for identification of mineral phases that cannot easily be identified under microscope. The following are only a few examples of minerals that can be successfully identified with the help of LRS.

Minerals in Bauxite and Laterite

Bauxite is a heterogeneous naturally occurring material with principal constituent minerals gibbsite ($Al_2O_3.3H_2O$), boehmite ($Al_2O_3.H_2O$ or γ-AlO(OH)) and diaspore (α-AlO(OH)), the later having the same composition as boehmite, but is denser and harder. These are mixed with the two iron oxides goethite and haematite, the clay mineral kaolinite and small amounts of anatase TiO_2. These phases are often difficult to identify separately in hand specimen or under microscope. EPMA cannot effectively separate them as they are all compositionally very similar. However, Laser Raman Spectroscopy, which depends on bond structure and is independent of mineral composition, has been shown to have distinct signatures for each of these phases and thus can be effectively used in phase identification.

Garnierite group of minerals, the Ni-Mg-bearing hydrous phyllosilicates found associated with Ni-laterite, include micron scale serpentine, talc, chlorite, smectite, sepiolite etc. Each of these phases can be identified by.

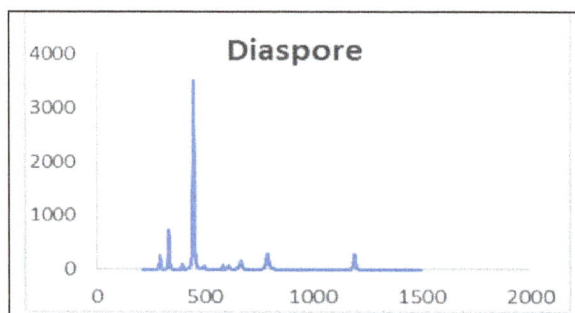

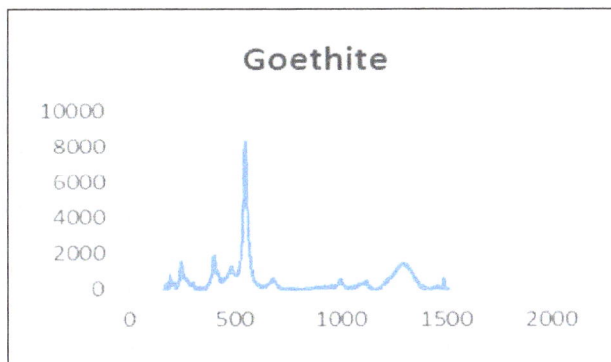

Iron Ore Minerals

The iron oxide minerals like magnetite (Fe_3O_4), hematite (Fe_2O_3), goethite(FeO(OH)), maghemite (Fe_2O_3, γ-Fe_2O_3), kenomagnetite (tetrahedral Fe_3O_4)are all structurally different and can be easily identified by their distinct LRS signatures.

Carbonate Phases

Carbonates provide a wealth of information about sedimentary environments and ancient climates, as well as being important in the formation of oil reservoirs; cement industry etc. The mineralogy of ooids is usually either calcite or aragonite, which correspond to climatic greenhouse and icehouse conditions: being able to distinguish the two is therefore important in interpreting the environmental significance of a particular rock formation.

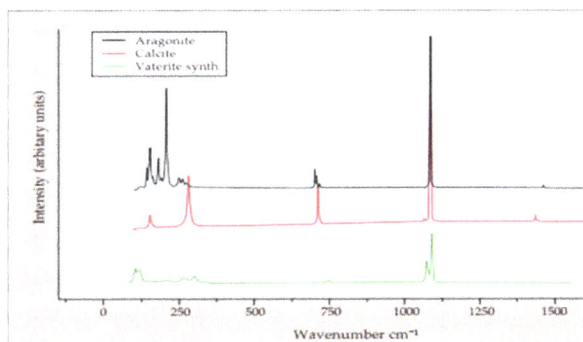

Carbonate minerals such as calcite, dolomite and aragonite have similar optical properties and distinguishing them is often difficult. Alizarin Red test can successfully demarcate some phases but fail to distinguish the others. Optical techniques also require the preparation of thin sections which is time-consuming and consequently expensive. Methods such as X-ray diffraction lack spatial resolution (in most cases being restricted to examination of crushed rock or single crystals extracted from a sample) and electron microscopy techniques require polished and carbon-coated surfaces and is still not useful for calcite and aragonite as these two phases are polymorphs and therefore cannot be distinguished by compositional data alone.

Laser Raman Spectroscope has distinct and well identified set of peak positions for each of these carbonate minerals and can hence be effectively applied for mineral identification in small limestone blocks and thin sections.

Sulphides

Sulfides (chalcopyrite, bornite, sphalerite, pyrite, marcasite, cubanite, pentlanadite, covellite, arsenopyrite) are weaker Raman scatters compared to silicates, carbonates and sulfate minerals. They have distinctive Raman bands in the ~300–500 $\Delta cm-1$ region. Raman spectra from these mineral species are very consistent in band position and normalized band intensity albeit rather overlapping. A few sulphides, such as galena and pyrrhotite, do not exhibit a first-order Raman spectrum due to their structural symmetry and Raman spectra could not be obtained from some of the metal-excess groups of sulphides because of their many metallic characteristics. However, good quality Raman spectra are possible from most sulphides and sulphosalts. However, the sulphide phases have distinct optical properties and can be easily identified through reflected light petrography.

Tin and Tungsten Minerals

The tin-tungsten minerals can be identified through petrographic study. But LRS provides an easier approach to mineral identification confirmation. The two wolframite end members ferberite (FeWO) and hubnerite (MnWO) have nearly identical structure and therefore identical peak positions in LRS. However studied signatures for scheelite (CaWO), stolzite (PbWO) and tungstite ($WO_8.H_2O$) are available and they vary sufficiently in details so as to allow LRS to identify these phases separately.

Cassiterite (SnO_2), stannite (Cu_2FeSnS_4), the rare mineral hemusite (Cu_6SnMoS_8) andkesterite ($Cu_2.o_4(Zno.84Feo.18)Sn1.01S_4$) have their distinct LRS signatures. However considerable degree of overlap between signatures of cernite (Cu_2CdSnS_4) and stannite make their distinction impossible under LRS.

Manganese Minerals

Manganese oxides, oxyhydroxides, carbonates and silicates constitute a large family of minerals. A few examples include manganosite (MnO), hausmannite ($Mn_2+Mn_3+2O_{4)}$, pyrolusite (MnO_2), bixbyite (Mn_2O_3), ramsdellite (Mn_4+O_2), jacobsite ($Mn_2+Fe_3+2O_4$), hollandite ($Ba(Mn_{4+6}Mn_{3+2})O_{16}$), manganite ($Mn_3+O(OH)$), allactite ($Mn_7(AsO_4)_2(OH)_8$), birnessite (Mn-oxyhydroxide), romanechite ($(Ba, H_2O)_2(Mn_4+ Mn_3+)5O_{10}$), braunite ($Mn_2+Mn_3+6SiO_{12}$), neltnerite ($CaMn_3+6O_8(SiO_4)$), rhodochrosite ($MnCO_3$) etc. These minerals have very similar optical properties which makes it extremely difficult to separate them through reflectance and transmission microscopy. Also due to large variations in oxidation states of manganese among the oxide minerals, identification of such phases with electron microprobe may not always be reliable. However, there exists extensive LRS data on the peak shifts depending on these mineral structures. Therefore LRS is the most reasonable and cost effective approach for manganese mineral identification.

Manganese Nodules/Polymetallic Nodules

Polymetallic nodules, also called manganese nodules were discovered at the end of the 19th century in the Kara Sea, in the Arctic Ocean off Siberia. During the scientific expeditions of the H.M.S. Challenger, they were found to occur in most oceans of the world. The important constituents of polymetallic nodules include todorokite [(Na,Ca,K,Ba,Sr)1-x(Mn,Mg,Al)6O12·3-4H2O], whichis a complex hydrous manganese oxide mineral,vernadite [(Mn$_4$+,Fe$_3$+,Ca,Na)(O,OH)$_2$nH$_2$O] and buserite (Na$_4$Mn14O27·21H$_2$O). These phases are frequently present as alternate micron scale discontinuous lamellae. LRS can prove be a method for identification of such constituent phases as each of these minerals have separate set of Stoke's shift values.

Aluminosilicate Polymorphs

The aluminosilicate polymorphs, kyanite, sillimanite and alusite have distinct optical properties and crystal habits. Thus easily recognisable in transmitted light microscopy. However, as it often happens in medium to low grade pelites, later alteration or retrograde metamorphism often leads to partial to complete replacement of the aluminosilicate phases. At times, only patches of the original minerals are preserved within lower grade phyllosilicate aggregates. Under such circumstances, identification of the aluminosilicate concerned becomes difficult with straight forward petrographic approach. But proper identification of the Al$_2$SiO$_5$ polymorph is absolutely essential both for understanding the geothermobarometric conditions and for determination of beneficiation processes. EPMA study is no alternative since the concerned phases are polymorphs and have same chemical formula. However, LRS gives distinct signatures for each of these phases and can thus be used for their identification in problematic situations.

Carbon Allotropes

There are a wide variety of different carbon nanostructures, however they all have a few basic things in common. First, all of these materials are predominantly made up of pure carbon, and as such can be called carbon allotropes. The range of these materials starts with the well known allotropes of diamond and graphite, and continues on to encompass fullerenes, graphene and more complex structures such as carbon nanotubes that re finding increasingly important applications in a variety of industries. From a molecular perspective, these materials are all entirely composed of C-C bonds, although the orientation of these bonds is different in the different materials and therefore, to characterize their molecular structure in a meaningful manner, it is necessary to have a technique which is highly sensitive to even slight changes in orientation of C-C bonds.

For diamond, where the material consists of highly uniform C-C bonds in a tetrahedral crystal structure, the Raman spectrum is very simple. It consists of only a single band at 1332 cm^{-1} because all of the bonds in the crystal are of the same orientation and strength resulting in a single vibrational frequency. In contrast, the small crystal size of nanocrystalline diamond results in a finite-size effect in which the lattice is somewhat distorted. This is manifested in the Raman spectrum by a slightly down shifted tetrahedral sp^3 band. The additional band at 1620 cm^{-1} and the shoulders on the 1620 cm^{-1} and tetrahedral sp^3 band are also indicative of sp^2 bonded carbon that represents surface defect modes, which would be insignificant in larger diamond crystals. Finally, the very broad band around 500 cm^{-1} is indicative of some amorphous sp^3 bonded carbon.

The graphite spectrum has several bands in the spectrum and the main band has shifted from 1332 cm^{-1} in diamond to 1580-82 cm^{-1} in graphite. The reason for this is that graphite is composed of sp2 bonded carbon in planar sheets in which the bond energy of the sp2 bonds is higher than the sp3 bonds of diamond.

When comparing Raman spectra of graphene and graphite, at first glance the spectra look very similar. This is not too surprising as graphite is just stacked graphene. However, there are some significant differences, the most obvious being that the band at 2690-2700 cm-1, which is known as the G' band, is much more intense than the G band in graphene compared to graphite. You may have heard the G' band referred to as the 2D band; both 2D and G' are accepted names for this band.

The main feature in the C60 fullerenespectrum is a relatively sharp line at around 1462 cm-1, known as the pentagonal pinch mode. In contrast, the spectrum of C70 fullereneis littered with numerous bands. This is due to a reduction in molecular symmetry which results in more Raman bands being active.

Carbon nanotubes (SWCNT, DWCNT and their superset MWCNT) are cylindrical carbon tubes having distinct LRS signatures.

Gemstones

Raman spectroscopy is particularly useful in identifying and characterizing gemstones that cannot be subjected to cutting, polishing and EPMA studies. It has been used to distinguish between gemstone and its simulant, to identify gems that are set in jewellery / objects of historical importance, to make some estimation on gem composition from comparing acquired data with available experimental data, to identify inclusions within gems, to characterize fluid inclusions in gems and to understand nature of treatments in some treated gems. Some experimental works on possible differences in LRS signatures between natural and synthetic gems are being presently carried out.

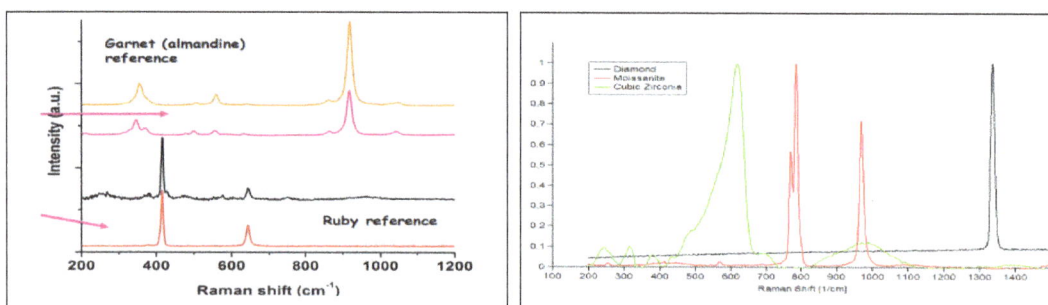

Identification of Vein Forming Fluids through Fluid Inclusion Study

Laser Raman Spectroscope can be ideally used, in conjunction with heating cooling stage for fluid inclusion study, to identify and characterize the fluid inclusions in minerals. Fluid inclusions are microscopic entrapments of liquid and gas within crystals. As ore minerals often form from a liquid or in an aqueous medium, tiny blebs of that liquid can become trapped within the crystal structure or in healed fractures within a crystal. These small inclusions range in size from 0.1 to 1 mm and are usually only visible in detail by microscopic study. Fluids play an important role in geological processes and fluid inclusion study helps in understanding of ore genetic processes.

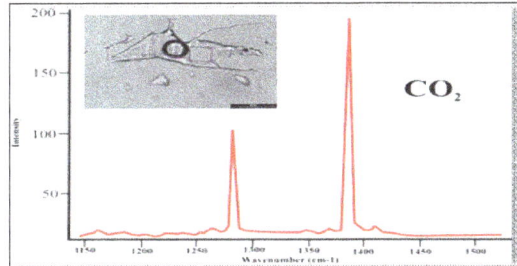

Laser Raman Spectroscope can penetrate through a transparent crystal/wafer into the trapped fluid inclusion. Such inclusions can be monophase, biphase or polyphase. The individual phases within the inclusion can be identified by focussing the laser beam on each of them separately and receiving the corresponding signatures. The knowledge about the nature of ore forming fluids and the trapped salts and resultant salinity can be used successfully for ore genetic interpretation.

Identification of Inclusions in Transparent Gem Variety Minerals

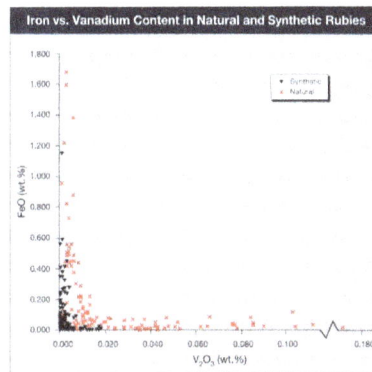

The LRS is a completely non-destructive process. The laser beam can penetrate a transparent mineral without causing any harm to its structure or composition. Thus the laser is used to penetrate gem variety minerals/crystals and collect spectrum from mineral inclusions inside the gemstone. Such data can be used for mineral inclusion identification from inside gemstones which on its turn helps in understanding of gem genesis and therefore further gemstone prospecting.

Shocked Minerals

LRS peak position, FWHM, symmetricity etc. depend on molecular vibration, which on its turn, is affected by lattice deformation. Thus LRS has been successfully used for study of shock metamorphic polymorphs in meteorites and impact craters.

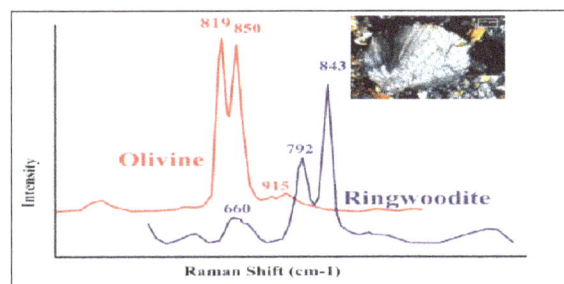

IR Laser Spectroscopy

IR laser spectroscopy has been employed to study free radicals and ions because of its sensitivity in the detection of compounds present in small concentrations. The spectroscopic characterization of radicals and ions is valuable in the fields of chemical kinetics, of astronomy and of plasma diagnosis. The experimental techniques used to produce unstable molecules and to record the spectra are here briefly summarized. Free radicals can be generated using the discharge-flow method, while ionic species can be obtained using a hollow-cathode discharge cell that, combined with magnetic-field modulation or amplitude-discharge modulation, allows the detection of the selected ionic species among neutral stable molecules. Alternatively, velocity modulation IR laser spectroscopy can be used to selectively detect ionic species among neutral ones, exploiting the characteristic Doppler shift of charged molecules. HD^+, HeH^+, NeH^+, ArH^+, HCO^+ and HNN^+ are some of the ions studied with this technique in the infrared and whose fundamental frequencies and molecular constants have been determined.

Both radicals and ions may be generated by photolysis or photoionization of a stable precursor seeded in a rare gas free-jet expansion. Another powerful technique with which to study ions employs a fast ion beam extracted from a plasma ion source and velocity tuned. This is merged with CO_2 laser light to induce transitions from a vibrational bound level near the dissociation limit to a repulsive electronic state or to a predissociating bound level above the dissociation limit. Most of the systems studied are diatomic, such as HD^+, HeH^+, He_2^+, H_2^+ and D_2^+, and the information obtained concerns the description and spectroscopy of weakly bound states, to a infer deeper understanding of ion–neutral reaction dynamics and of the charge–dipole induced interaction. In addition, polyatomic unstable species have been the subject of detailed study with high-resolution IR spectroscopy, among those examined are CCH, C_3, C_4, C_9, C_{13}, NO_3, HO_2, HCCO and C_3H_3 radicals, and H_3^+, SiH_3^+, CH_3CNH^+, CH_2^+, NH_3^+ and H_2O^+ ions. The results of such experiments may be valuable in the field of chemical kinetics since they allow the determination of rate constants and an understanding of some aspects of the mechanism of the reaction. For example, the observation, in a hollow cathode discharge, of the v_1 band of HOC^+ at 3268 cm;$^{-1}$ allowed the measurement of the abundance ratio $[HCO^+]:[HOC^+]$ in the laboratory from the relative intensity of the IR high-resolution lines of the two species, using the transition dipole moments for the band of HCO^+ (0.168 D) and HOC^+ (0.350–D), and assuming that the rotational temperature of both ions was equal to the cell temperature (223–K) (Amano). This information was particularly valuable in studying the formation and depletion of HOC^+ in the interstellar medium.

A rate equation analysis was performed to rationalize the observed dependence of the relative intensities of HOC^+ and HCO^+ lines as a function of the pressures of H_2 and CO. The rate constants of some of the reactions considered in the rate equation analysis were then determined. The contribution that the analysis of the high-resolution IR spectra of such kind of molecules brings to astronomy is the determination with great accuracy of the molecular constants that allow one to search for the rotational lines in laboratory microwave spectra. These results may lead to the identification of unassigned interstellar lines. The nature and the amount of a species present in plasmas may be determined by high-resolution IR spectroscopy, as in the case of SiH_3^+, a transient molecular ion observed and analysed in a silane discharge plasma. The v_2 and v_4 bands, i.e. the out-of-plane bending and the in-plane degenerate bending vibration respectively, have been recorded and analysed taking into account the Coriolis interaction between the two vibrationally excited states.

Ultrafast Laser Spectroscopy

Ultrafast laser spectroscopy involves studying ultrafast events that take place in a medium using ultrashort pulses and delays for time resolution.It usually involves exciting the medium with one (or more) ultrashort laser pulse(s) and probing it a variable delay later with another.

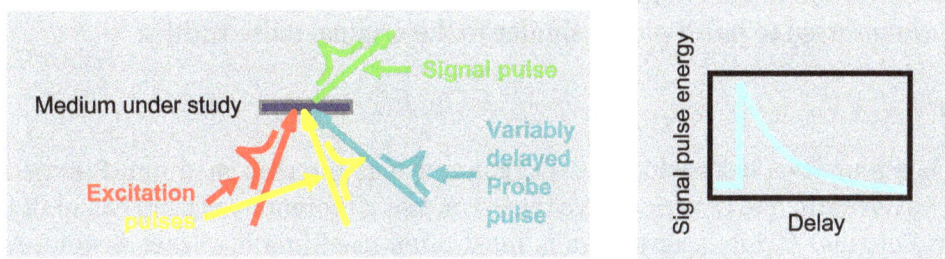

The signal pulse energy (or change in energy) is plotted vs. delay.

The experimental temporal resolution is the pulse length.

Spectroscopy Measurements

The excite pulse(s) excite(s) molecules into excited states, which changes the medium's absorption coefficient and refractive index.

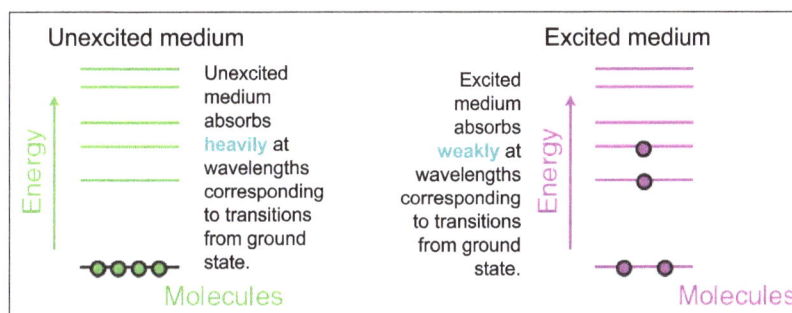

The excited states only live for a finite time (this lifetime is often the quantity we'd like to find), so the absorption and refractive index return to their initial values eventually.

Attosecond-to-picosecond Spectroscopy

Dynamics on the as to fs time scale is in general too fast to be measured electronically. Most measurements are done by employing a sequence of ultrashort light pulses to initiate a process and record its dynamics. The width of the light pulses have to be on the same scale as the dynamics that is to be measured.

Light Sources

Titanium-sapphire Laser

Ti-sapphire lasers are tunable lasers which emit red and near-infrared light (700 nm- 1100 nm).

Ti-sapphire laser oscillators use Ti doped-sapphire crystals as a gain medium and Kerr lens modelocking to achieve sub-picosecond light pulses. Typical Ti-sapphire oscillator pulses have nJ energy and repetition rates 70-100 MHz. Chirped pulse amplification through regenerative amplification can be used to attain higher pulse energies. To amplify, laser pulses from the Ti-sapphire oscillator must first be stretched in time to prevent, damage to optics, and then are injected into the cavity of another laser where pulses are amplified at a lower repetition rate. Regeneratively amplified pulses can be further amplified in a multi-pass amplified. Following amplification, the pulses are recompressed to pulse widths similar to the original pulse widths.

Dye Laser

A dye laser is a four-level laser which uses organic dye as the gain medium. Pumped by a laser with a fixed wavelength, due to various dye types you use, different dye lasers can emit beams with different wavelengths. A ring laser design is most often used in a dye laser system. Also, tuning elements, such as a diffraction grating or prism, are usually incorporated in the cavity. This allows only light in a very narrow frequency range to resonate in the cavity and be emitted as laser emission. The wide tuneability range, high output power, and pulsed or CW operation make the dye laser particularly useful in many physical & chemical studies.

Fiber Laser

A fiber laser is usually generated first from a laser diode. The laser diode then couples the light into a fiber where it will be confined. Different wavelengths can be achieved with the use of doped fiber. The pump light from the laser diode will excite a state in the doped fiber which can then drop in energy causing a specific wavelength to be emitted. This wavelength may be different from that of the pump light and more useful for a particular experiment.

X-ray Generation

Ultrafast optical pulses can be used to generate x-ray pulses in multiple ways. An optical pulse can excite an electron pulse via the photoelectric effect, and acceleration across a high potential gives the electrons kinetic energy. When the electrons hit a target they generate both characteristic x-rays and bremsstrahlung. A second method is via laser induced plasma. When very high intensity laser light is incident on a target, it strips electrons off the target creating a negatively charged plasma cloud. The strong Coulomb force due to the ionized material in the center of the cloud quickly accelerates the electrons back to towards the nuclei left behind. Upon collision with the nuclei, Bremsstrahlung and characteristic emission x-rays are given off. This method of x-ray generation scatters photons in all directions, but also generates picosecond x-ray pulses.

Conversion and Characterization

Pulse Characterization

In order for accurate spectroscopic measurements to be made, several characteristics of the laser pulse need to be known; pulse duration, pulse energy, spectral phase and spectral shape are among some of these. Information about pulse duration can be determined through autocorrelation measurements, or from cross correlation with another well characterized pulse. Methods allowing for complete characterization of pulses include frequency-resolved optical gating (FROG) and spectral phase interferometry for direct electric-field reconstruction (SPIDER).

Pulse Shaping

Pulse shaping is to modify the pulses from the source in a well-defined manner, including manipulation on pulse's amplitude, phase and duration. To amplify pulse's intensity, chirped pulse amplification is generally applied, which includes a pulse stretcher, amplifier and compressor. It will not change the duration or phase of the pulse during the amplification. Pulse compression (shorten the pulse duration) is achieved by first chirping the pulse in a nonlinear material and broadening the spectrum, with a following compressor for chirp compensation. Fiber compressor is generally used in this case. Pulse shapers usually refer to optical modulators which applies Fourier transforms to laser beam. Depending on which property of light is controlled, modulators are called intensity modulators, phase modulators, polarization modulators, spatial light modulators. Depending on the modulation mechanism, optical modulators are divided into Acoustic-optic modulators, Electro-optic modulators, Liquid crystal modulators etc. Each is dedicated into different applications.

High Harmonic Generation

High harmonic generation (HHG) is the nonlinear process where intense laser radiation is converted from one fixed frequency to high harmonics of that frequency by ionization and recollision of an electron. It was first observed in 1987 by McPherson et al. who successfully generated harmonic emission up to the 17th order at 248 nm in neon gas. HHG is seen by focusing an ultra-fast, high-intensity, near-IR pulse into a noble gas at intensities of (10^{13}–10^{14} W/cm^2) and it generates coherent pulses in the XUV to Soft X-ray (100–1 nm) region of the spectrum. It is realizable on a laboratory scale (table-top systems) as opposed to large free electron-laser facilities.

High harmonic generation in atoms is well understood in terms of the three-step model (ionization, propagation and recombination):

- Ionization: The intense laser field modifies the Coulomb potential of the atom, electron tunnels through the barrier and ionize.

- Propagation: The free electron accelerates in the laser field and gains momentum.

- Recombination: When the field reverses, the electron is accelerated back toward the ionic parent and releases a photon with very high energy.

Frequency Conversion Techniques

Different spectroscopy experiments require different excitation or probe wavelengths. For this reason frequency conversion techniques are commonly used to extend the operational spectrum of existing laser light sources. The most widespread conversion techniques rely on using crystals with second order non linearity to perform either parametric amplification or frequency mixing. Frequency mixing works by superimposing two beams of equal or different wavelengths to generate a signal which is a higher harmonic or the sum frequency of the first two. Parametric amplification overlaps a weak probe beam with a higher energy pump beam in a non linear crystal such that the weak beam gets amplified and the remaining energy goes out as a new beam called the idler.

This approach has the capability of generating output pulses that are shorter than the input ones. Different schemes of this approach have been implemented. Examples are: optical parametric oscillator (OPO), optical parametric amplifier (OPA), non-collinear parametric amplifier (NOPA).

Techniques

Ultra-fast Transient Absorption

This method is typical of 'pulse-probe' experiments, where a pulsed laser is used to excite a molecule's electrons from their ground states to higher-energy excited states. A probing light source, typically a xenon arc lamp, is used to obtain an absorption spectrum of the compound at various times following its excitation. As the excited molecules absorb the probe light, they are further excited to even higher states. After passing through the sample, the unabsorbed light from the arc lamp continues to an avalanche photodiode array, and the data is processed to generate an absorption spectrum of the excited state. Since all the molecules in the sample will not undergo the same dynamics simultaneously, this experiment must be carried out many times, and the data must be averaged in order to generate spectra with accurate intensities and peaks. Unlike TCSPC, this technique can be carried out on non-fluorescent samples.

Ultrafast transient absorption can use almost any probe light, so long as the probe is of a pertinent wavelength or set of wavelengths. A monochromator and photomultiplier tube in place of the avalanche photodiode array, allows observation of a single probe wavelength, and thus allows probing of the decay kinetics of the excited species. The purpose of this setup is to take kinetic measurements of species that are otherwise nonradiative, and specifically it is useful for observing species that have short-lived and non-phosphorescent populations within the triplet manifold as part of their decay path. The pulsed laser in this setup is used both as a primary excitation source, and a clock signal for the ultrafast measurements. Although laborious and time-consuming, the monochromator position may also be shifted to allow absorbance decay profiles to be constructed, ultimately to the same effect as the above method.

Time-resolved Photoelectron Spectroscopy and Two-photon Photoelectron spectroscopy

Time-resolved photo-electron spectroscopy and two-photon photoelectron spectroscopy (2PPE) combine a pump-probe scheme with angle-resolved photoemission. A first laser pulse is used to excite a material, a second laser pulse ionizes the system. The kinetic energy of the electrons from this process are then detected, through various methods including energy mapping, time of flight measurements etc. As above, the process is repeated many times, with different time delays between the probe pulse and the pump pulse. This builds up a picture of how the molecule relaxes over time. A variation of this method looks at the positive ions created in this process, and is called time-resolved photo-ion spectroscopy (TRPIS).

Multidimensional Spectroscopy

Using the same principles pioneered by 2D-NMR experiments, multidimensional optical or infrared spectroscopy is possible using ultrafast pulses. Different frequencies can probe various dynamic molecular processes to differentiate between inhomogeneous and homogeneous line broadening as well as identify coupling between the measured spectroscopic transitions. If two oscillators are coupled together, be it intramolecular vibrations or intermolecular electronic coupling, the added dimensionality will resolve anharmonic responses not identifiable in linear spectra. A typical 2D pulse sequence consists of an initial pulse to pump the system into coherent superposition of states, followed by a phase conjugate second pulse that pushes the system into a non-oscillating excited state, and finally, a third pulse that converts back to a coherent state that produces a measurable pulse. A 2D frequency spectrum can then be recorded by plotting the Fourier transform of the delay between the first and second pulses on one axis, and the Fourier transform of the delay between a detection pulse relative to the signal-producing third pulse on the other axis. 2D spectroscopy is an example of a four wave mixing experiment, and the wavevector of the signal will be the sum of the three incident wavevectors used in the pulse sequence. Multidimensional spectroscopies exist in infrared and visible variants as well as combinations using different wavelength regions.

Ultrafast Imaging

Most ultrafast imaging techniques are variations on standard pump-probe experiments. Some commonly used techniques are Electron Diffraction imaging, Kerr Gated Microscopy, imaging with ultrafast electron pulses and terahertz imaging. This is particularly true in the biomedical community where safe and non-invasive techniques for diagnosis are always of interest. Terahertz imaging has recently been used to identify areas of decay in tooth enamel and image the layers of the skin. Additionally it has shown to be able to successfully distinguish a region of breast carcinoma from healthy tissue. Another technique called Serial Time-encoded amplified microscopy has shown to have the capability of even earlier detection of trace amount of cancer cells in the blood. Other non-biomedical applications include ultrafast imaging around corners or through opaque objects.

Femtosecond up-conversion

Femtosecond up-conversion is a pump-probe technique that uses nonlinear optics to combine the fluorescence signal and probe signal to create a signal with a new frequency via photon upconversion, which is subsequently detected. The probe scans through delay times after the pump excites the sample, generating a plot of intensity over time.

Applications

Applications of Femtosecond Spectroscopy to Biochemistry

Ultrafast processes are found throughout biology. Until the advent of femtosecond methods, many of the mechanism of such processes were unknown. Examples of these include the cis-trans photoisomerization of the rhodopsin chromophoreretinal, excited state and population dynamics of DNA, and the charge transfer processes in photosynthetic reaction centers. Charge transfer dynamics in photosynthetic reaction centers has a direct bearing on man's ability to develop light harvesting technology, while the excited state dynamics of DNA has implications in diseases such as skin cancer. Advances in femtosecond methods are crucial to the understanding of ultrafast phenomena in nature.

Photodissociation and Femtosecond Probing

Photodissociation is a chemical reaction in which a chemical compound is broken down by photons. It is defined as the interaction of one or more photons with one target molecule. Any photon with sufficient energy can affect the chemical bonds of a chemical compound, such as visible light, ultraviolet light, x-rays and gamma rays. The technique of probing chemical reactions has been successfully applied to unimolecular dissociations. The possibility of using a femtosecond technique to study bimolecular reactions at the individual collision level is complicated by the difficulties of spatial and temporal synchronization. One way to overcome this problem is through the use of Van der Waals complexes of weakly bound molecular cluster. Femtosecond techniques are not limited to the observation of the chemical reactions, but can even exploited to influence the course of the reaction. This can open new relaxation channels or increase the yield of certain reaction products.

Picosecond-to-Nanosecond Spectroscopy

Streak Camera

Unlike attosecond and femtosecond pulses, the duration of pulses on the nanosecond timescale are slow enough to be measured through electronic means. Streak cameras translate the temporal profile of pulses into that of a spatial profile; that is, photons that arrive on the detector at different times arrive at different locations on the detector.

Time-correlated Single Photon Counting

Time-correlated single photon counting (TCSPC) is used to analyze the relaxation of molecules from an excited state to a lower energy state. Since various molecules in a sample will emit photons at different times following their simultaneous excitation, the decay must be thought of as having a certain rate rather than occurring at a specific time after excitation. By observing how long individual molecules take to emit their photons, and then combining all these data points, an intensity vs. time graph can be generated that displays the exponential decay curve typical to these processes. However, it is difficult to simultaneously monitor multiple molecules. Instead, individual excitation-relaxation events are recorded and then averaged to generate the curve.

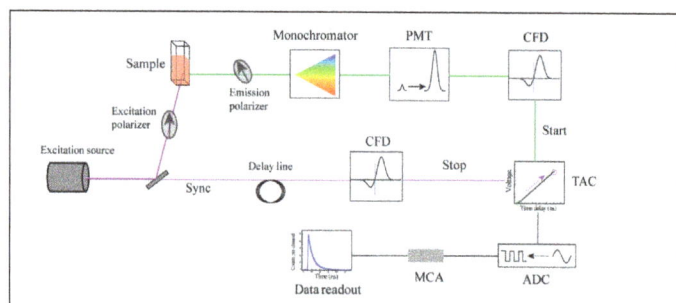

This technique analyzes the time difference between the excitation of the sample molecule and the release of energy as another photon. Repeating this process many times will give a decay profile. Pulsed lasers or LEDs can be used as a source of excitation. Part of the light passes through the sample, the other to the electronics as "sync" signal. The light emitted by the sample molecule is passed through a monochromator to select a specific wavelength. The light then is detected and amplified by a photomultiplier tube (PMT). The emitted light signal as well as reference light signal is processed through a constant fraction discriminator (CFD) which eliminates timing jitter. After passing through the CFD, the reference pulse activates a time-to-amplitude converter (TAC)

circuit. The TAC charges a capacitor which will hold the signal until the next electrical pulse. In reverse TAC mode the signal of "sync" stops the TAC. This data is then further processed by an analog to digital converter (ADC) and multi-channel analyzer (MCA) to get a data output. To make sure that the decay is not biased to early arriving photons, the photon count rate is kept low (usually less than 1% of excitation rate).

This electrical pulse comes after the second laser pulse excites the molecule to a higher energy state, and a photon is eventually emitted from a single molecule upon returning to its original state. Thus, the longer a molecule takes to emit a photon, the higher the voltage of the resulting pulse. The central concept of this technique is that only a single photon is needed to discharge the capacitor. Thus, this experiment must be repeated many times to gather the full range of delays between excitation and emission of a photon. After each trial, a pre-calibrated computer converts the voltage sent out by the TAC into a time and records the event in a histogram of time since excitation. Since the probability that no molecule will have relaxed decreases with time, a decay curve emerges that can then be analyzed to find out the decay rate of the event.

A major complicating factor is that many decay processes involve multiple energy states, and thus multiple rate constants. Though non-linear least squared analysis can usually detect the different rate constants, determining the processes involved is often very difficult and requires the combination of multiple ultra-fast techniques. Even more complicating is the presence of inter-system crossing and other non-radiative processes in a molecule. A limiting factor of this technique is that it is limited to studying energy states that result in fluorescent decay. The technique can also be used to study relaxation of electrons from the conduction band to the valence band in semiconductors.

Frequency Comb Spectroscopy

A laser frequency combs is a broad spectrum composed of equidistant narrow lines. Initially invented for frequency metrology, such combs enable new approaches to spectroscopy over broad spectral bandwidths, of particular relevance to molecules. With optical frequency combs, the performance of existing spectrometers, such as Michelson-based Fourier transform interferometers or crossed dispersers, involving e.g. virtual imaging phase array (VIPA) étalons, is dramatically enhanced. Novel types of instruments, such as dual-comb spectrometers, lead to a new class of devices without moving parts for accurate measurements over broad spectral ranges. The direct selfcalibration of the frequency scale of the spectra within the accuracy of an atomic clock and the negligible contribution of the instrumental line-shape will enable determinations of all spectral parameters with high accuracy for stringent comparisons with theories in atomic and molecular physics. Chip-scale frequency-comb spectrometers promise integrated devices for real-time sensing in analytical chemistry and biomedicine.

A frequency comb is a spectrum of phase-coherent evenly spaced narrow laser lines. Frequency combs1 have revolutionized time and frequency metrology in the late 1990's by providing rulers in frequency space that measure large optical frequency differences and/or straightforwardly link microwave and optical frequencies. Very rapidly, frequency combs have found applications beyond the original purpose. For instance, they provide long-term calibration of large astronomical spectrographs; by enabling the control of the relative phase between the envelope and the carrier of ultrashort pulses, they have become a key to attosecond science; low-noise frequency combs of high repetition frequency benefit radio-frequency arbitrary waveform generation and optical communications. In the present review article, we focus on their impact in spectroscopy where the frequency comb is used to directly excite or interrogate the sample. This field is sometimes called direct frequency comb spectroscopy, or broadband spectroscopy with frequency combs. In the following, we coin it frequency comb spectroscopy.

The advent of frequency comb spectroscopy brings a set of new tools to spectroscopy in all phases of matter. In the simplest approach, a frequency comb excites and interrogates the sample. The spectral response of the sample, due e.g. to linear absorption or to a nonlinear phenomenon, may span the entire bandwidth of the comb and therefore a spectrometer is required. Despite daunting technical challenges, the last decade has witnessed remarkable progress in laser frequency-comb sources dedicated to broadband spectroscopy, especially in the molecular-fingerprint mid-infrared (2-20 μm) region and the ultraviolet range ($< 400\,nm$). Existing spectrometers and spectrometric techniques are adapted and improved to resolve individual comb lines, while comb-enabled instruments are explored. First studies suggest new opportunities for exploring atomic and molecular structure and dynamics. Meanwhile, spectroscopic sensors, with enhanced capabilities to probe a variety of environments, are devised.

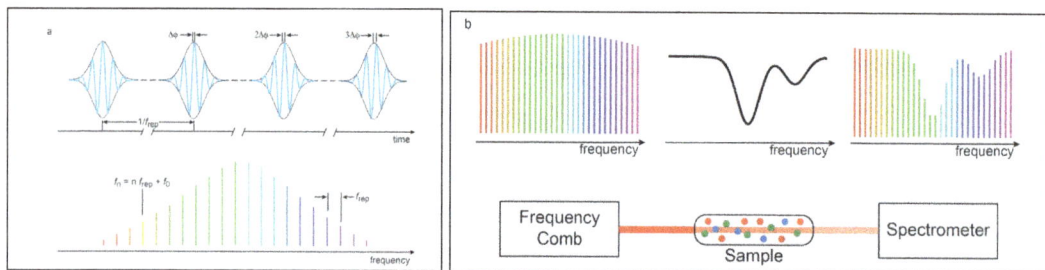

Principle of a frequency comb and sketch of a simple experiment of frequency comb spectroscopy. a. Time domain representation of the train of ultra-short pulses of period $1/f_{rep}$ at the output of a mode-locked laser and the corresponding spectrum of narrow lines of a frequency comb. The phase-shift $\Delta\phi$ of the carrier of the wave relatively to the envelope of the pulses induces a translation $\Delta\phi\, f_0 = f_{rep}\, \Delta\phi / 2\pi$ of all the lines in the spectrum from their harmonic frequencies nf_{rep}. b. In the simplest experiment of frequency comb spectroscopy, a frequency comb as a broadband light source interrogates an absorbing sample and a spectrometer analyzes the transmission spectrum.

Frequency-comb Light Sources for Spectroscopy

Characteristics

Often, a comb is generated by a mode-locked laser system. In the time domain, a train of ultrashort pulses is emitted at the output of the cavity. The period of the envelope of the pulses, $1/f_{rep} = L/v_g$, corresponds to the round-trip time inside a laser cavity of round-trip length L and group velocity of the light vg. Due to dispersion inside the cavity, a pulse-to-pulse phase-shift $\Delta\phi$ between the carrier and the envelope of the electric field of the pulses is observed. In the frequency domain, the associated spectrum is composed of a discrete set of evenly spaced narrow lines with frequencies f_n that can be written $f_n = n f_{rep} + f_0$, where n is a large integer, f_{rep} is the repetition frequency of the envelope of the pulses and f_o is the carrier-envelope offset frequency, related to the phase-shift $\Delta\phi$ by the relationship $f_0 = f_{rep} \Delta\phi / 2\pi$.

All regions of the electro-magnetic spectrum offer interesting opportunities for Spectroscopy. However, many applications involve interrogating strong rotational or ro-vibrational transitions in molecules or electronic (rovibronic) transitions in atoms (molecules). Therefore, the targeted regions are the submillimetre (also called THz. Wavelength range: 100μm-1mm; frequency range: 0.3-3THz), the far- (100-20 μm, 3-15THz), the mid- infrared (2-20 μm, 15- 150THz) and the ultra-violet ($< 400\,nm$, > 750 THz) ranges.

A broad spectral span is advantageous, especially for spectroscopy of molecules. Most spectrometers have a limited dynamic range and therefore combs with a flat intensity spectral distribution are also beneficial. Furthermore, not all applications require self-referencing. Many combs for frequency comb spectroscopy use the output of the laser system, possibly moderately broadened in a nonlinear fibre, rather than the octave-spanning spectrum, used for instance in f-2f interferometers, that suffers from strong variations of intensity across the span.

Experiments of linear spectroscopy usually do not require a large average power, as the single photodiode or the camera of the spectrometer can only admit a limited power. Therefore, many experiments of absorption spectroscopy harness an average power on the μW- or mW-scale. Conversely, experiments involving the generation of nonlinear phenomena at the sample often require high pulse energies and therefore higher average power, up to the W level, are reported.

The most suited repetition frequency strongly depends on the type of spectroscopy and the type of sample. Ideally, the repetition frequency should be similar to the desired spectral resolution. Furthermore, resolving the individual comb lines brings the possibility of defining the spectral elements more precisely than by their spacing and of directly calibrating the frequency scale, rather than relying on spectral reference lines. For samples in the condensed phase, a line spacing in the range 50-500 GHz may be desirable. Conversely, for the study of Doppler-broadened transitions of light gas molecules at room temperature in the mid-infrared region, a repetition frequency in the range 50-200 MHz is advantageous. Doppler-free spectroscopy, heavy molecules, or cold samples may require an even narrower spacing. However, few spectrometers have a sufficient resolution to resolve such comb lines directly. The figure below provides an illustration of the spans and repetition frequencies of selected comb generators across the sub-millimetre and infrared regions.

Overview of the spectral spans and repetition frequencies accessible with femtosecond lasers (bar in black), semiconductor lasers (red), microresonator based Kerr combs (dark blue), difference frequency generation (violet), optical parametric oscillators (green) and electro-optic modulators (cyan) from the THz to the near-infrared region. The width of each bar represents the spectral span over which the respective comb emits (dashed line: tuning range, solid line: span for one setting) and the ordinate gives the repetition frequency of the source. The selection, which focuses on comb sources designed for spectroscopy, is intended as an illustration and it does not provide an exhaustive summary of available frequency comb generators.

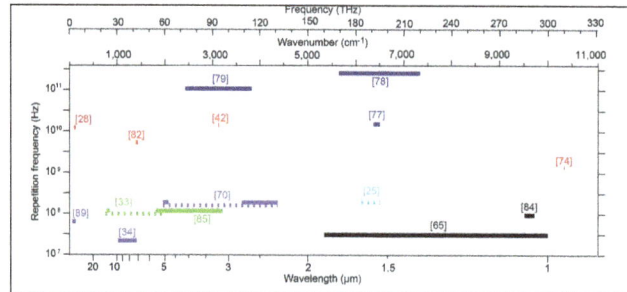

Figure: Spectral coverage with a selection of sources available for frequency comb spectroscopy.

Near-infrared Region

Frequency-comb technology in the near-infrared region (800nm–2μm, 150-375 THz) is now mature. Frequency-comb mode-locked lasers are commercially available, the most widespread gain media being Ti:Sa and fibres doped with ytterbium, erbium, thulium or holmium. Recent developments of sources without mode-locked lasers expand the capabilities in terms of repetition frequency, span, compactness etc. Many approaches that have a potential for operation in the mid-infrared region are first tested in the telecommunication region around 1.5 μm (200 THz). Soliton Kerr combs generated in high-quality factor micro-resonators, provide convenient access to a range of repetition frequencies, from 10 GHz to 1 THz, which is difficult, or even impossible, to reach with conventional mode-locked lasers. The recent demonstration of a battery-operated Kerr frequency-comb generator, as well as that of a III-V-on-Si mode-locked laser, holds much promise for fully integrated chip-based synthesizers. Frequency-agile combs, based on architectures involving one or several electro-optic modulators, exhibit flat-top spectra with a moderate number of lines (up to 10,000) but a freely selectable line spacing and a centre frequency that can be rapidly tuned.

Sub-millimetre (THz) and Far-infrared Region

Frequency-comb sources directly emitting in the THz (1 mm-100 μm, 0.3-3 THz) and far-infrared (100-20 μm, 3-15 THz) regions have long remained inexistent. Therefore, the main approach for generating a THz frequency comb has been to down-convert a near-infrared frequency comb by nonlinear frequency conversion using photomixing in photoconductive antennas or optical rectification in crystals. Offset-free combs, of an average power usually limited to a few microwatts, are produced with the same line spacing as the near-infrared combs. The use of fibre lasers as near-infrared pumps and the design of novel photoconductive emitters with plasmonic enhancement point to THz sources of smaller footprint and higher average power.

THz quantum-cascade frequency-comb lasers hold promise for versatile and compact high-power electrically-pumped semiconductor sources that do not produce pulses, but still generate an array of a few hundreds of phase-coherent lines by four-wave mixing. With a careful design of the dispersion compensation, frequency-comb operation over a bandwidth close to an octave, around a central frequency of 3 THz (100μm), features a power of 10 mW and a comb-line spacing of about 13 GHz.

Conversely, at large-scale facilities, synchrotron radiation in the so-called coherent mode shows temporal coherence, from one electron bunch to the next and over a revolution period in the storage ring. This intriguing phenomenon remains to be exploited for spectroscopy.

Mid-infrared Region

Similarly, the mid-infrared region (2-20 µm, 15-150 THz) is technologically challenging. Ref. reviews the field until 2012. Since then, direct mid-infrared generation of ultra-short pulses has progressed with e.g. a promising modelocked Er3+ doped fluoride glass fibre laser of 200-fs pulse duration at 2.8µm (110 THz). As erbium, thulium and holmium present several gain bands in the mid-infrared region, further breakthroughs may be expected. Remarkable advances have also been achieved with new materials and platforms for nonlinear frequency conversion and with quantum cascade lasers. Access to long wavelengths up to 12 µm (frequencies down to 25 THz) has become possible with nonlinear crystals such as orientation-patterned gallium phosphide or $LiGaS_2$. Microresonators and waveguides generate or broaden midinfrared spectra, with on-chip silicon or silicon-nitride platforms, as reviewed in Ref. Quantum-cascade and interband-cascade 42 laser frequency combs of GHz line spacing cover several spectral bands between 3 and 9 µm (33-100 THz) at an average power up to the Watt level, with about 100-200 comb lines. As quantum cascade devices also operate as detectors, they show an intriguing potential for fully integrated broadband spectrometers.

Visible and Ultraviolet Regions

The visible and near-ultraviolet regions, down to wavelengths of about 200 nm (1,500 THz), are conveniently accessible by frequency-comb techniques using spectral broadening of near-infrared lasers in nonlinear waveguides and/or harmonic- or sum-frequency conversion in nonlinear crystals. New crystals, such as $KBe_2BO_3F_2$ or $Li_4Sr(BO_3)_2$, may extend the range down to 160 nm (up to 1,870 THz). Reaching shorter wavelengths involves high-harmonic generation in a rare gas, a very non-efficient process. For generating frequency combs of high repetition frequency (>20 MHz) suited for direct frequency comb spectroscopy in the extreme ultraviolet, the approach has been to inject the equidistant modes of an infrared, or visible, frequency comb into a resonant passive cavity containing the focus for the gas target. After each pass through the focus, the non-converted portion of the light pulse is coherently overlapped with the successive pulse from the laser. In this way, the intensity enhancement needed for high-harmonic generation can be reached. The approach is complex, as it also requires suitable out-coupling of the ultraviolet light and optimization of phasematching effects that control the build-up of the harmonic signal over the interaction length. Recently, a record average power of 0.7 mW for a harmonic at 63 nm (4,760 THz) has been reported 46 at a repetition frequency of 77 MHz, using a swept cavity.

Spectrometric Techniques for Frequency Comb Spectroscopy

In most cases, the frequency-comb generator is a broadband light source that simultaneously excites several (many) transitions of the sample. Therefore, a spectrometer is needed, except for limited comb spans and/or very simple spectra. If the spectrometer has sufficient resolution, the individual comb lines may be resolved, enabling self-calibration of the frequency scale. Then the resolution is determined by the comb repetition frequency f_{rep}, although the spectral elements may be defined with a significantly higher precision. Once the resolution of the spectrometer is equal to – or better than – the comb line spacing, the instrumental line-shape, that convolves the atomic or molecular transitions, becomes determined by the width of the individual comb lines rather than by the spectrometer response. With a comb of narrow lines, the contribution of the instrumental line-shape becomes negligible when the atomic or molecular resonances have a width similar to or broader than the line spacing f_{rep} of the comb. Before the advent of frequency comb spectroscopy, the instrumental line-shape in multiplex or multichannel spectroscopy used to be, at best, of similar width as the transitions. Furthermore, assuming that the sample does not change with time, the resolution may be enhanced, fundamentally down to the intrinsic comb line-width, by interlaying several spectra recorded e.g. with a tuned comb offset frequency f_0.

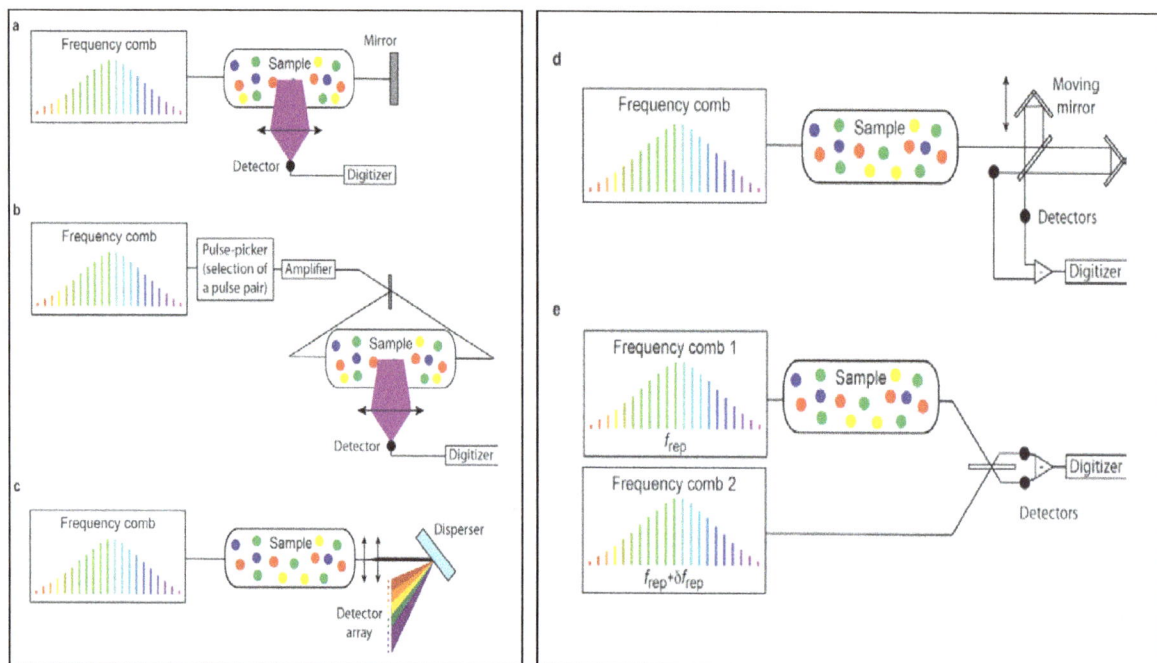

Spectrometric techniques for frequency comb spectroscopy:

a. Direct frequency comb spectroscopy with the example of two-photon Doppler-free excitation in a standing wave and detection of fluorescence of the sample.

b. Ramsey-comb spectroscopy also with the example of two-photon Doppler-free excitation in a standing wave and detection of fluorescence of the sample.

c. Frequency-comb spectrometry with a disperser for absorption measurements. Here a simple grating and a detector array are represented.

d. Frequency-comb Fourier transform spectroscopy with a scanning Michelson interferometer and an absorbing sample.

e. Dual-comb spectroscopy with one comb interrogating the sample and the other acting as a local oscillator. The absorption and the dispersion of the sample are measured.

Direct Frequency Comb Spectroscopy

Direct frequency comb spectroscopy is the simplest approach to linear or nonlinear frequency comb spectroscopy. For linear spectroscopy, a single comb line is resonant with a transition and all the other lines should ideally be detuned from resonances. For two-photon excitation, many pairs of comb lines of the same sum frequency contribute to the excitation, which can be as efficient as with a continuous-wave laser of the same average power. The excitation can even be Doppler-free if the atoms are excited by two counterpropagating pulses forming a standing wave. The different schemes of twophoton excitation include stimulated Raman effects. The response of the sample, e.g. its transmission, its fluorescence or its ionization rate, is recorded using a single detector. Sweeping e.g. the comb carrier-envelope frequency f0 scans the spectrum, which is measured with a free-spectral range equal to the comb line spacing f_{rep}, ideally large. The approach is therefore only suitable for simple spectra comprising a few narrow transitions within the range of excitation and it has consequently been limited to gas-phase atomic systems. The technique is powerful, though: it is simple to implement compared to techniques requiring a spectrometer; absolute frequency calibration is obtained through knowledge of the repetition frequency f_{rep} and of the carrier-envelope offset frequency f_0; as frequency combs often involve intense ultra-short pulses, nonlinear frequency conversion may be efficient and allows interrogating transitions in spectral ranges that are difficult or impossible to access with continuous-wave lasers. Pulse shaping allows to reduce the residual Doppler effect and to excite different transitions at distinct spatial locations. For nonlinear excitation in the vacuum ultraviolet, reaching sufficient power with high-harmonic comb generators of large line spacing is a major challenge.

Ramsey-comb Spectroscopy

Ramsey-comb spectroscopy is a related time-domain technique that measures the interference between the excitations of an atomic or molecular sample by two time-delayed intense pulses derived from a frequency comb. The fringes, which are sensitive to the phase of the second pulse relative to the atomic coherence, are sampled at a set of different delay times, which are integer multiples of the pulse spacing of the laser oscillator plus some chosen fractional increments. The frequency of the excited transitions is deduced from fits of the portions of the phase signals. Similarly to direct frequency comb spectroscopy, the free spectral is limited to the comb repetition frequency f_{rep}, so that the technique is mostly suitable for metrology of simple spectra with few transitions. As already demonstrated in the deep ultraviolet with two-photon transitions of H_2 around 202 nm (1,485 THz), Ramsey-comb spectroscopy holds particular promise, because the pairs of phase-coherent infrared pulses amplified to the millijoule level allow efficient frequency conversion.

Spectroscopy using a Dispersive Spectrometer

Dispersive spectrographs provide simple and robust tools for multichannel approaches to broadband spectroscopy with frequency combs. Gratings 15 and crossed dispersers, utilizing e.g. virtual imaging phase array (VIPA) étalons, have been successfully implemented with scanning single detectors and with cameras. With a crossed disperser, resolutions as high as 600 MHz have been reported. Most of the time, this is insufficient to resolve individual comb lines and many reports do not rely on the calibration by the frequency comb. Low-resolution spectrographs are however sufficient to resolve the individual lines of chip-based frequency comb generators such as microresonators. As advanced control and tuning of the line positions per steps across one free spectral range is feasible, rapid and compact instruments for gas-phase spectroscopy can even be envisioned. Alternatively, Fabry-Pérot filter cavities can increase the comb line spacing to exceed the spectrograph resolution. Vernier techniques, where the engineered mismatch between the free spectral range of a scanning Fabry-Pérot cavity and the frequency-comb line spacing is chosen as a ratio m/(m-1) with m integer, considerably relax the constraints on the spectrograph resolution. The Fabry-Pérot resonators may present a high finesse of several thousands: therefore they may simultaneously serve as enhancement cavities for weakly absorbing samples. Thanks to the availability of mid-infrared cameras with a short integration time, cavityenhanced frequency-comb spectroscopy with a VIPA has proven a powerful tool for monitoring, with a time resolution of 10 μs, the kinetics of the gas-phase reaction between carbon monoxide and the hydroxyl radical and for observing the intermediate hydrocarboxyl radical.

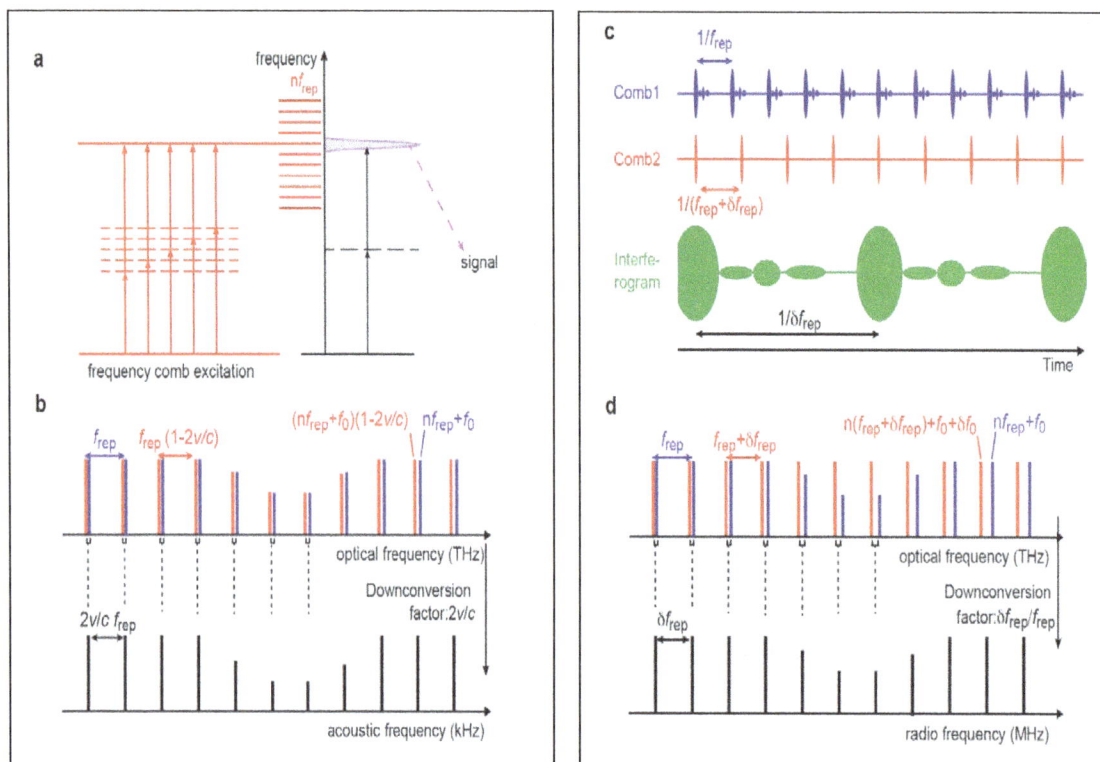

Physical principle of some of the described spectrometric techniques:

a. In direct frequency comb spectroscopy with two-photon excitation, many pairs of comb lines may contribute to the excitation of the transition. However the spectrum is only measured modulo the comb repetition frequency. Fluorescence during decays towards lower energy levels may be detected.

b. In the moving arm of a scanning Michelson interferometer, the frequency of all the comb lines is Dopplershifted. The beat notes between pairs of shifted and unshifted comb lines at the detector produce an acoustic comb.

c. Interferometric sampling in the time-domain stretches freeinduction decay. With a dual-comb system, the interferogram recurs automatically at a period $1/\delta f_{rep}$ which is the inverse of the difference in repetititon frequencies of the two combs.

d. Frequency-domain picture of c for dual-comb interferometry. The beat notes between pairs of comb lines, one from each comb, generates a radio-frequency comb. The physical principle is the same as that of b., except that the down-conversion factor no longer depends on the speed of a moving part. Furthermore, dualcomb systems render the implementation of a dispersive interferometer easier.

Michelson-based Fourier Transform Spectroscopy

Fourier transform spectroscopy with a scanning Michelson interferometer has been one of the most successful spectrometric techniques over the past fifty years. Usually associated with an in-coherent broadband light source, the spectrometer measures on a single photo-detector the inter-ference between the two optically delayed signal from the two arms as a function of the optical path difference. The spectrum is the Fourier transform of the time-domain interference waveform, the interferogram.

Fourier transform spectrometry makes the best use of the available time and photons. It records spectra over extended spectral spans in any spectral regions and the spectral data, all simultane-ously recorded, exhibit quality and consistency. Its instrumental line-shape is well understood and modelled. Its limitations are related to the presence of moving parts: the resolution is inversely proportional to the excursion of the moving arm and high-resolution instruments, commercially available with resolutions as high as 30 MHz, are slow and bulky.

In a Michelson interferometer, the frequency f of the light traveling in the moving arm acquires a small Doppler-shift, equal to -2 f v/c, where v is the velocity of the mirror, c is the speed of light. At the output of the interferometer, the two electric fields coming from the fixed and moving arms beat on a photo-detector. The detector signal comprises this interference pattern, resulting from the down-conversion of the optical frequencies mostly to the audio-range. When a frequency comb is used as a light source in front of the interferometer, the frequency of each comb line, reflected in the moving arm of the interferometer, is shifted by a factor - $2\left(n f_{rep} + f_0\right)v/c$, which gives the frequencies of the acoustic comb generated at the detector. The use of the frequency-comb synthe-sizer 19 has shown to add significant improvements to Fourier transform spectroscopy. A coherent light source such as a laser frequency comb has significantly higher brightness, leading to increased signalto-noise ratio or decreased measurement times. Resolving the comb lines brings instrumental line-shapes of negligible contribution and direct calibration of the frequency scale (assuming that the comb is self-referenced), whereas accurate line position measurements previously relied on the presence of reference lines and careful assessment of the systematic effects. The frequency comb enables the implementation of detection techniques that straightforwardly retrieve the dispersion spectrum, leading to the measurement of real and imaginary part of the refractive index. The de-velopment of the technique has been fast, as it has profited from existing advanced techniques of control of scanning interferometers. Its use has been so far restricted to gas-phase spectroscopy in the infrared region. The sensitivity has been enhanced by synchronous detection 19, multi-pass cells 19 or high finesse cavities. Cavity-enhanced frequency-comb Fourier transform spectroscopy has shown remarkable results for disentangling complex spectra of heavy molecules.

Dual-comb Spectroscopy

Dual-comb spectroscopy is a comb-enabled approach to Fourier transform interferometry without moving parts. This instrumental scheme of frequency comb spectroscopy is currently that which attracts the highest interest. In most of the implementations, a frequency comb, of repetition fre-quency f_{rep}, interrogates the sample and beats on a fast photodiode with a second comb, of slightly different repetition frequency $f_{rep} + \delta f_{rep}$, which acts as a local oscillator. The interference signal is recorded as a function of time and it is Fourier transformed to reveal the spectrum. In the time domain, the pulses from one comb walk through the pulses of the second comb with a time delay that automatically increases of an amount $\delta f_{rep}/f_{rep2}$ from pulse pair to pulse pair. After interacting with the sample, the repeating waveforms of the electric field of the first comb are optically sampled by the local-oscillator comb and provide an interferogram stretched in time by a factor $f_{rep} + \delta f_{rep}$. In this way, optical delays between 0 and $1/f_{rep}$ are periodically scanned. In the frequency domain, the beating signal of the two optical frequency combs of slightly different line spacing produces a comb in the radio-frequency domain that can directly be measured by digital electronics.

The physical principle of the measurement is the same as that of the scanning Michelson-based Fourier transform spectrometer, with the practical difference that the down-conversion factor is not set by the speed of a moving part. It is therefore freely selectable within the Nyquist limit. The dual-comb spectrum may thus be mapped into the radio-frequency region, where the 1/f noise is reduced. A Michelson interferometer leaves to the user the choice of the resolution lower than the instrument capabilities: it allows any path difference excursions shorter than its maximum path difference, whereas the dual-comb spectrometer, by construction, automatically scans the optical delays up to $1/f_{rep}$. Even if numerical treatments, called apodization, make it possible to reduce the resolution, the time-domain scan always reaches a spectral resolution equal to the comb line spacing f_{rep}. If the desired resolution is significantly lower than f_{rep}, this results in wasted experimental time. Therefore, for optimized measurement times, dual-comb systems with comb line spacing of the same order of magnitude as the desired resolution are advantageous in most cases. Conversely, the resolution of spectra with resolved comb lines can always be improved by interleaving spectra.

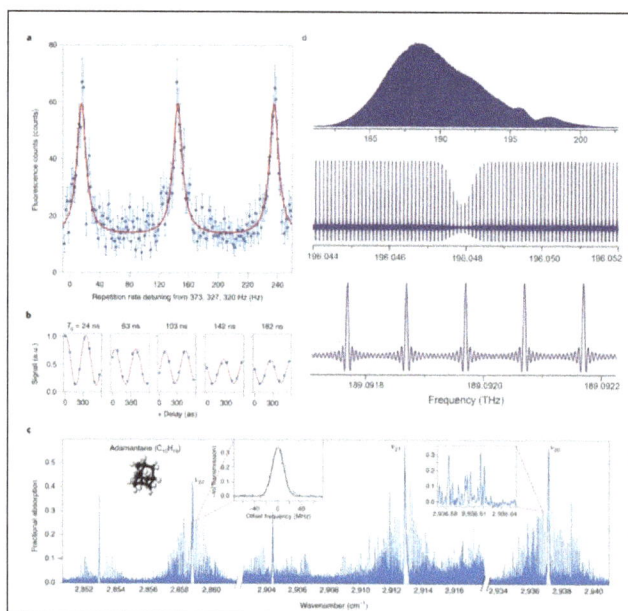

Illustration of experimental results from the different approaches to frequency comb spectroscopy:

a. Direct frequency-comb one-photon spectrum 61 of the D_2 line of a single $25Mg^+$ ion around 280 nm (1,070 THz) observed through fluorescence. The spectral line recurs, as the free-spectral range is about 373 MHz.

b. Ramsey-comb interference fringes of the $EF^1\Sigma_g^+ - X1\Sigma_g^+$ $(0,0)$ $Q1$ rovibronic Doppler-free transition of H_2 excited by two photons at 202 nm (1,485 THz).

c. Cavity-enhanced frequency-comb Fourier transform spectrum of buffer-gas cooled ada-mantine ($C_{10}H_{16}$) in the 3-μm region, at a translational temperature of 17 K and a spectral resolution of 10 MHz.

d. Near-infrared experimental dual-comb spectrum showing 200,000 resolved comb lines with the expected cardinal-sine instrumental line-shape. The mutual coherence time of the interferometer is close to 2,000 seconds.

A fundamental difference between dual-comb spectrometers and traditional ones is that the du-al-comb approach is freed from geometry. With a dispersive or interferential instrument, the the-oretical resolving power R may be expressed as $R = \Delta / \lambda$, the ratio of the maximum path differ-ence Δ to the wavelength λ. With a dual-comb interferometer, it becomes $R = T / \tau$ where T is the measurement time and τ the period of the light vibration. Dual-comb spectroscopy is therefore the only technique that can, for any spans and any spacing, potentially reach a resolution equal to the comb line spacing. Such a specific feature may open up novel opportunities in precision spectroscopy and metrology, although it has not been exploited yet because of the technical chal-lenges.

In addition to the challenges associated with the availability of the laser sources in the spectral regions of interest, a specific difficulty has slowed the applications of dual-comb spectroscopy to spectroscopic studies. The field has long been dominated by preliminary proofs-of-principle with a variety of original laser systems. The efficient measurement of interference requires time-domain coherence between the two interfering electric fields. Small relative timing- and phase-fluctuations between the two combs appear stretched by the same factor as the optical-ly sampled waveform. In the frequency domain, the width of the beat notes between pairs of comb lines, one from each comb, needs to be narrower than the spacing of the radio-frequency comb lines, δf_{rep}, to preserve resolution. Furthermore, their intrinsic width should be narrow-er than the inverse of the measurement time, for best signal-to-noise ratio. The constraint is the same as that which require the interferometric control of the path difference in a Michel-son-based Fourier transform spectrometer, technically mastered for decades. In dual-comb interferometry, the powerful yet simple techniques for referencing the comb to a radio-fre-quency clock, which are widely adopted in frequency metrology, do not provide the required short-term relative stability.

One approach has been to experimentally achieve such a mutual coherence by sophisticated servo controls. Locking two lines of each comb to a pair of narrow line-width continuous-wave lasers has yielded a coherence time inversely proportional to the line-width of the continuous-wave lasers. Using this principle, continuous time-domain measurements on the order of 1 second can be per-formed 64, 65. Numerical techniques known as phase correction, derived from Michelson-based Fourier transform spectroscopy, can then be applied to efficiently average many recordings of 1 second and improve the signal-to-noise ratio. More recently, by feed-forward relative stabilization of the carrierenvelope offset frequency, the experimental coherence time reaches 2,000s, with-out any indications that a limit is reached, in the near-infrared and in the mid-infrared regions. This suggests that, as with a Michelson interferometer, the experimental phase control of the du-al-comb interferometer can be arbitrarily long, opening up novel opportunities for broadband fre-quency metrology.

Another approach has been to track the relative fluctuations between the two combs and to correct for these, either in real-time, by analog or digital processing, or a posteriori. Such schemes have been implemented even with free-running or loosely locked lasers. They also compensate for the residual fluctuations of stabilized systems.

A third trend, which is currently stimulating many creative experiments, is to design dual-comb systems with built-in passive mutual coherence. Two trains of asynchronous pulses may be generated in dual-wavelength unidirectional modelocked lasers, in bidirectional mode-locked lasers and micro-resonators e.g. by taking advantage of an asymmetry to nonlinear processes. A bire fringent plate in a laser cavity has been harnessed to produce overlapping crossed polarized pulse trains of a difference in repetition frequencies that depends on the optical thickness of the plate. The two combs of electro-optic-modulator based systems can share a number of components. With some of these set-up, the mutual-coherence time exceeds 1 second. Such designs without any stabilization electronics greatly facilitate dual-comb spectroscopy and hold promise for transportable or even portable compact and easy-to-use interferometers.

Most of the time, dual-comb interferometers are designed for linear absorption spectroscopy. When the sample only interacts with one comb, the phase spectrum is simultaneously obtained, whereas it was technically challenging to investigate with dispersive Michelson interferometers. Direct access to both the real and imaginary parts of the refractive indices is given. With the objective of linear absorption measurements, numerous demonstrations have been accomplished with a variety of laser sources, mostly across the infrared region; to cite just a few: near-infrared erbium-doped and ytterbium-doped fibre combs, electro-optic modulators, near-infrared and mid-infrared microresonators, frequency-doubled fibre lasers mid-infrared and THz quantum cascade or interband cascade lasers, mid-infrared Cr^{2+}:ZnSe lasers mid-infrared optical parametric oscillators mid-infrared difference-frequency systems THz photoconductive antennas.

As dual-comb interferometers often harness ultra-short-pulse lasers, they enable novel nonlinear broadband spectroscopy. New schemes of coherent Raman spectroscopy (CARS), stimulated Raman spectroscopy and two-photon excitation have been first demonstrated, followed by others such as pumpprobe spectroscopy. Sometimes, e.g. in dual-comb two-photon excitation with background-free fluorescence detection, the interferometric modulation is measured indirectly, through its transfer to e.g. the modulation of the intensity of the fluorescence of the sample, providing an insightful illustration of Fourier encoding. The first proof-of-concept of broadband Doppler-free two-photon spectroscopy showcases experimental spectra spanning 10 THz, which exhibit atomic line profiles of a width of 6 MHz.

A strength of dual-comb spectroscopy (and other techniques of Fourier transform spectroscopy), that has been recently explored, is its ability to efficiently combine with other sampling techniques or instrumentations. The sensitivity to weak absorption is enhanced with multipass cells and enhancement cavities the signal-to-noise ratio is increased by electro-optic sampling spectro-imaging and microscopy provide spectral maps of spatially inhomogeneous samples. The attractive advantage of multiplex measurements over extended spectral spans, which provides overall consistency and short measurement times, is added to those of the sampling techniques. Specific features of dualcomb systems include the absence of moving parts, the use of laser beams rather than incoherent light, the feasibility of short measurement times, the absolute frequency calibration and the achievable negligible contribution of the instrumental line-shape.

Other Approaches

Alternate methods, involving e.g. speckles in multimode fibres sweeping the comb repetition frequency to generate a time delay between the two arms of a static interferometer cavity filtering and scanning or heterodyning a comb with a continuous-wave laser 100 have been explored. Though they have not been widely adopted yet, they may be particularly useful in some circumstances.

Selected Applications and Prospects

Precision Spectroscopy

Precision spectroscopy of atomic transitions has been the most investigated application of frequency comb spectroscopy. Such measurements enable stringent tests of fundamental theories, accurate determinations of physical constants, and searches for new physics. With direct frequency comb spectroscopy or Ramsey-comb spectroscopy, the absolute frequency of narrow transitions in atomic systems is determined, such as in argon, cesium, krypton, hydrogen, magnesium, single magnesium ion, neon, single calcium ion rubidium in gas cells, ion traps, laser-cooled systems or atomic beams. Only one molecule, H_2, has been considered so far. The fractional uncertainty of the frequency measurements is typically on the order of 10-10, but it can reach parts in 10^{12}. The extension of the techniques of precision spectroscopy with frequency-comb excitation to the extreme ultraviolet, where continuous-wave lasers are not available, portends fascinating opportunities for fundamental physics, including better tests of quantum electrodynamics, of molecular quantum theory, or future nuclear clocks.

Moreover, the prospect of Doppler-free spectroscopy or of cold-molecule spectroscopy over broad spectral bandwidths is opening new perspectives to precision spectroscopy: detailed analysis e.g. of simple molecules with few electrons over an extended range may deliver new information on molecular structure and potential-energy surface and may help to validate or improve ab initio quantum-chemical computations.

Laboratory Molecular Spectroscopy over Broad Spectral Bandwidths

Because of the instrumental challenges of frequency comb spectrometry, spectroscopic studies have remained rather scarce, but the few published ones are likely to stimulate further contributions from the growing community. Most of the spectroscopy work has been performed in the near-infrared region, where the technology is more mature. Mainly line positions and shifts have been determined. The unique feature of frequency-comb spectrometers with resolved narrow comb lines, the negligible contribution of the instrumental line-shape, will permit the metrology of line parameters other than line positions: frequency comb spectroscopy combines for the first time a broad spectral span, which was the distinctive feature of spectrometers with incoherent light sources, and an narrow instrumental line-width, which used to be the specific character of some tunable lasers. New investigations for a better understanding and modelling of spectral line-shapes may be triggered. One of the first published studies illustrates such benefits for the modelling of near-infrared spectra of water vapour at high temperature.

The combination of fast measurement times and (sometimes moderately) broad spectral bandwidth advances time-resolved spectroscopy, with time resolutions on the scale of several microseconds, in a variety of situations, from the investigation of chemical gas-phase reactions through mid-infrared spectroscopy to the kinetics of spectral hole burning in a transition of atomic caesium. Gas-phase transient absorption of electronic transitions of diatomic molecules in the visible range and vibrational and electronic population relaxation of dye molecules in solution 105 associate frequency comb spectroscopy to the study of ultrafast phenomena.

The past few years have witnessed a diversification of the samples, which are no longer restricted to the gas phase. Samples in the liquid or solid phases have been studied. With them comes the requirement of sources and spectrometric techniques suited to their broad spectral transitions and their extended spectra. The development of spectrometric techniques involving microresonator-based frequency combs of large line spacing, especially in the mid-infrared, is therefore very timely. Nonlinear spectroscopy with combs of high repetition frequency faces the specific difficulty of a lower energy per pulse and requires novel strategies to be devised.

Coherent Control and Multi-dimensional Spectroscopy

Laser frequency comb techniques can measure and control the phase of an optical electric field with respect to the corresponding intensity waveform. This is opening up new horizons for the generation of arbitrary waveforms at optical frequencies and "line-by-line" pulse shaping. It might create novel opportunities for coherent control in chemical reactions. The intense ultra-short laser pulses of the frequency combs can be further harnessed to exploit complex pulse sequences and coherent transient phenomena including photon echoes, in analogy to multidimensional nuclear magnetic resonance spectroscopy. On a longer term, by rapidly exploring a multi-dimensional parameter space, nonlinear multi-dimensional multi-comb spectroscopy and imaging might reveal much additional information inaccessible by conventional linear and nonlinear spectroscopy. Two-dimensional spectroscopy of gas-phase alkali atoms has been explored with a dual-comb 110 system generating photon echoes, at spectral resolutions that would be difficult to reach with the mechanical delay lines commonly used in multidimensional spectroscopy. Theoretical proposals and insights may help guiding the vision.

Environmental Sensing

Frequency comb spectroscopy presents some interesting characteristics for fieldable and even portable gas sensors. For instance, the laser beams enable long open-path propagation, filling the gap between point sensors and remotesensing instruments, e.g. on board satellites, aircrafts or balloons. The spectra show high consistency, stability and repeatability for concentration measurements. The broad spectral bandwidth enables detection of multiple species, as well as more reliable inversion. The applications range from the monitoring of greenhouse gases to industrial process control or leak detections. Significant progress has already been achieved toward the objective of compact portable in situ frequency-comb spectrometers. A transportable dual-comb sensor, deployed in the field, shows continuous monitoring and quantification of methane emission sources at a regional scale, with the prospect of efficient leak detection at oil and gas operation facilities. In a laboratory proof-of principle of dual-comb spectroscopy of laser-induced plasmas, time-resolved broadband spectral analysis of laser-ablated solid materials is performed and lays the first bases for in situ laser-induced breakdown spectroscopy of solids, liquids and aerosols. The demonstrations have so far been accomplished in the near infrared region. Improvements to fibre-, semiconductor- and chip-based instrumentation will render the sensors more compact, rugged and easy-to-use, even in harsh environments. Continued progress to mid-infrared frequencycomb sensing technology will increase the number of detectable molecules, as well as the detection sensitivity.

Applications to Chemistry, Biology and Medicine

Frequency combs will expand the capabilities of optical spectroscopy, spectromicroscopy and hyperspectral imaging for chemical or bio-medical analysis. Breath analysis by cavity-enhanced direct frequency-comb spectroscopy has also been envisioned. An even more intriguing prospect is the potential of frequency combs for physical chemistry in condensed matter. Indeed, harnessing frequency combs for "low-resolution" spectroscopy may initially be seen as nonintuitive and thought provoking. However, converging insights and first proof-ofconcepts provide a set of arguments. Chip-scale dual-comb spectrometers with mid-infrared combs of large line-spacing may bring new tools for timeresolved spectroscopy of samples in the condensed phase. First dual-comb spectrometers for spectro-imaging, confocal microscopy 96 and near-field microscopy showcase a short measurement time per pixel and a high spectral resolution.

Photoacoustic Spectroscopy

Photoacoustic spectroscopy (PAS) is based on the absorption of electromagnetic radiation by analyte molecules. Non-radiative relaxation processes (such as collisions with other molecules) lead to local warming of the sample matrix. Pressure fluctuations are then generated by thermal expansion, which can be detected in the form of acoustic or ultrasonic waves. In other words, the transformation of an optical event to an acoustic one takes place in photoacoustic spectroscopy. A fraction of the radiation falling upon the sample is absorbed and results in excitation, the type of which being dependent upon the energy of the incident radiation. Non-radiative de-excitation (relaxation) processes which normally occur give rise to the generation of thermal energy within the sample. If the incident radiation is modulated then the generation of thermal energy within the sample will also be periodic and a thermal wave/pressure wave will be produced having the same frequency as this modulation. Energy is transferred by the thermal wave/pressure wave towards the sample boundary, where a periodic temperature change is generated. The periodic variation in the temperature at the surface of the sample results in the generation of an acoustic wave in the gas immediately adjacent and this wave propagates through the volume of the gas to the detector (microphone, piezoelectric transducers or optical method) where a signal is produced. This detector or microphone signal, when plotted as a function of wavelength, will give a spectrum proportional to the absorption spectrum of the sample. Therefore, the photoacoustic signal is a function of two types of processes occurring in the sample: the absorption of electromagnetic radiation specified by the absorption coefficient β and the thermal propagation in the sample specified by the thermal diffusivity, x.

The photoacoustic effect was discovered by A.G. Bell. He found that thin discs emit sound when exposed to a rapidly interrupted beam of sunlight. By placing different absorbing substances in contact with the ear using a hearing tube, he was able to detect absorption in both the visible and the invisible regions of the solar spectrum. This spectrophone was used in his experiments on wireless transmission of sound. Important steps leading to this rediscovery of the effect for analytical purposes were the invention of the laser as an intense light source, the development of highly sensitive sound detectors (such as condenser microphones and piezoelectric transducers), and the first comprehensive theoretical description of the photoacoustic effect in solids by Rosencwaig and Gersho: the so-called RG theory.

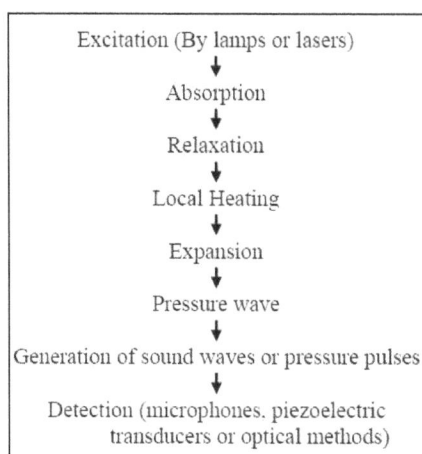

Figure: Principle of excitation, signal generation and detection in a photoacoustic experiment.

Techniques Involved in Photoacoustic Spectroscopy

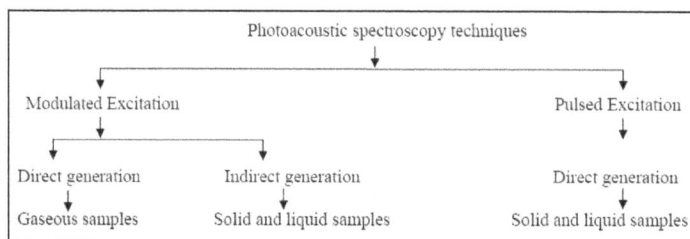

Excitation

The Photoacoustic effect is based on the sample heating produced by optical absorption. In order to generate acoustic waves, which can be detected by pressure sensitive transducers, periodic heating and cooling of the sample is necessary to generate pressure fluctuations. In principle, there are two ways to realize Photoacoustic pressure fluctuations: modulated and pulsed excitation.

Modulated Excitation

In modulated excitation schemes, radiation sources are employed whose intensity fluctuates periodically in the form of a square or a sine wave, resulting in a 50% duty cycle. This can be realized for example by the mechanical chopping of a light source. A way to overcome the 50% duty cycle is to modulate the phase instead of the amplitude of the emitted radiation. Whereas chopped or modulated lamps or IR sources from commercial spectrometers are used for the determination of UV/Vis or IR absorption spectra of opaque solids, modulated continuous wave (cw) lasers are the most common sources for Photoacostic gas phase analysis. The modulation frequencies usually range from a few Hz up to several kHz. The resulting pressure fluctuations generate sound waves in the audible range, which can be detected by microphones in the gas phase. Data analysis is performed in the frequency domain. Therefore, lock-in amplifiers are used for signal recording, which allows the analysis of both the amplitude and the phase of the sound wave. Since the acoustic propagation during the relatively long illumination period is much larger than the dimensions of the sample in most cases, boundary conditions have to be taken into account. This means that eigenmodes of the PA cell play an important role. This fact can be utilized for signal enhancement by acoustic resonance. Thus, acoustic resonance curves must be considered in PA cell design. In solid samples, tuning of the modulation frequency allows depthresolved investigations.

Pulsed Excitation

In pulsed PAS, laser pulses with durations in the nanosecond range are usually employed for excitation. Since the repetition rates are in the range of a few Hz, the result is a short illumination followed by a much longer dark period: a low duty cycle. This leads to a fast and adiabatic thermal expansion of the sample medium resulting in a short shock pulse. Data analysis in this case is performed in the time domain. Therefore, the signal is recorded by oscilloscopes, boxcar systems, or fast A/D converters. Transformation of the signal pulse into the frequency domain results in a wide spectrum of acoustic frequencies up to the ultrasonic range. Thus, laser beams modulated in the form of a sine wave excite one single acoustic frequency, whereas short laser pulses are broadband acoustic sources. In solid samples, analysis of the time delay between laser pulse and pressure detection allows depth profiling.

Signal Generation

Induction of an acoustic wave by modulated or pulsed irradiation inside a gaseous, liquid or solid sample is termed as direct Photoacostic generation. Here, detection takes place inside or at an interface of the sample. In indirect Photoacostic generation, heat is generated by modulated illumination inside a solid or liquid sample and transported to an interface. Subsequently, sound waves are generated and detected in the gas phase adjacent to the sample.

Direct PA Generation in Gases

Chopped continuous wave lasers or modulated laser diodes are employed for the modulated excitation of PA signals in gases. The modulated laser beam irradiates the gaseous sample inside a (usually cylindrical) PA cell, and sound waves with an acoustic frequency defined by the modulation frequency of the laser can be detected using microphones. The signal amplitude can be described by, $p = FW_0\alpha$, where W_0 is the incident radiation power and a is the absorption coefficient of the sample. The proportionality factor F is termed the cell constant. In the so-called non-resonant mode, the modulation frequency is much lower than the first acoustic resonance frequency of the photoacoustic cell. In this case, the wavelength of the generated acoustic wave is larger than the cell dimensions. Thus, the generation of standing acoustic waves is not possible. The cell constant F is, $F = G(\gamma-1)L/\omega V$. Here, G is a geometric factor of the order of one, γ is the adiabatic coefficient of the gas, L and V are the length and the volume of the cell, and $\omega = 2\pi\nu$ the modulation frequency.

The photoacoustic signal is indirectly proportional to the modulation frequency and the crosssection V/L of the cell. Thus, the signal increases with decreasing cell dimensions and modulation frequency. As the noise increases with a decrease in these parameters, there is a maximum in the S/N ratio for a certain combination of cell size and modulation frequency.

With increasing modulation frequency, at a certain point the acoustic wavelength reaches the cell dimensions, and the resonant eigenmodes of the cell can be excited, leading to an amplification of the signal. Resonance properties mainly depend upon the geometry and size of the cavity. The modulation frequency is tuned to one of the eigen frequencies of the cell and thus the signal is amplified by a quality factor Qi resulting in a cell constant $F_{res} = QC_i(\gamma - 1)/\omega_i V$, where Ci denotes a factor that depends on the positions of the laser beam and the microphone relative to the pressure distribution in the cell. The index i indicates that Q and C are parameters describing the eigenmode of the cell.

The quality factor can be described as the ratio between the accumulated energy in the resonator and the energy loss per cycle. In resonant cells with high quality factors, the resonance profile becomes narrow and small drifts of modulation frequency or speed of sound in the gas mixture cause strong changes in Q. Therefore, moderate Q-factors are preferred in many applications. As mentioned, pulsed excitation leads to broadband generation of various acoustic frequencies. Thus, in gas phase analysis by pulsed laser PAS, all eigenmodes of the cell can be excited simultaneously. The cell design is usually optimized to the excitation of one selected eigenmode, which is well separated from the neighboring ones. In this case, each laser pulse generates an exponentially decaying sine wave whose amplitude can be described by, $p = C(\gamma - 1)E_0 \mu_a / V$. Here, E_0 is the laser pulse energy. Note that the PA signal amplitude does not depend on Q-factor, modulation frequency or repetition rate. Thus, the signal does not depend on the speed of sound in the cell. In this case, the cell constant depends only on the geometrical properties of the PA cell. If the signal is dominated by one eigenmode and the contributions of other eigenmodes can be neglected, cell constants can be calculated that are in good agreement with experimental results. Pulsed laser PAS can therefore be applied to absolute absorption measurements in gases.

Direct PA Generation in Liquids and Solids

In condensed matter, short laser pulses are used for direct PA generation. The short illumination with relatively high peak power leads to an instantaneous adiabatic expansion of the medium, generating pressure pulses that propagate through the sample at the speed of sound. These ultrasonic pulses can be detected directly at a boundary of the sample by piezoelectric transducers or optical methods. As in other PA techniques, the signal amplitude depends linearly on the excitation energy and the absorption coefficient of the sample, and they can be described by, $P \propto \beta c^2 E_0 \mu_a / C_p$ where β, c and C_p denote the samples thermal expansion coefficient, speed of sound, and heat capacity, respectively. If the signal in pulsed PAS is recorded time-resolved, the time delay t between laser pulse and pressure detection can be determined, which represents the propagation time of the ultrasonic pulse through the sample. Thus, the depth z of an absorbing object inside the sample can be calculated simply as follows $z = ct$. This equation corresponds to the principle of ultrasonic tomography. It should be pointed out that in ultrasonic tomography; reflections of pressure waves at acoustic impedance mismatches are detected. Thus, for a signal to occur in ultrasonic tomography, a change of the acoustic impedance Z at an interface is needed, which is defined as $Z = \rho$, where ρ is the density of the sample. In contrast to ultrasonic tomography, PA depth profiling allows the depth resolved detection of changes in optical absorption. The maximum depth that can be investigated by pulsed PAS is limited by the optical penetration depth, $\delta = 1/\mu a$. Since the decay length of acoustic waves is much higher than the optical penetration depth in many solid samples, beyond this limit depth profiling is possible by detecting laser-induced pressure pulses, which are reflected at acoustic impedance mismatches. This technique, which is similar to ultrasonic tomography, can be applied to the nondestructive testing of opaque solid materials, and is termed laser-induced ultrasound (LIU).

Depth resolution of both pulsed PAS and LIU depends mainly on the time resolution of the ultrasonic detector. The depth resolution can be calculated as the product of the temporal resolution of the detector and the speed of sound in the sample. If fast piezoelectric detectors and data recording with temporal resolutions in the nanosecond range are used, depth resolutions in the lower micrometer range can be realized. The maximum sampling depth can reach a few centimeters in weak absorbing and scattering samples. If piezoelectric detector arrays, scanning PA sensors or suitable optical methods for detection are employed, two dimensional and three dimensional imaging are feasible by pulsed PAS.

Indirect PA Generation

Analysis of condensed matter by modulated PA excitation and subsequent detection of the directly generated acoustic wave by a microphone is not suitable due to strong acoustic impedance mismatches between solid and gas phase. Thus, an indirect scheme for PA generation is employed. Modulated warming of the sample is induced by modulated excitation. Subsequently, the heat deposited in the sample is transported to the interface of the sample with the adjacent gas phase. This heat transport can be described as thermal wave.

For indirect generation and detection of PA signals, PA cells can be coupled to conventional spectrometers, allowing UV/Vis or IR absorption spectroscopy of opaque solid samples. A relatively common set-up for this purpose is FT-IR–PAS. In this case, a Fourier transform IR spectrometer is used for excitation and PA signals are detected in a PA cell, which is in contact with the sample. In continuous scan FT-IR (CSFT-IR), the modulation frequency can be tuned by changing the interferometer moving mirror velocity. In step scan mode (S^2FT-IR), the excitation beam is modulated mechanically by a chopper for intensity modulation or by jittering the position of one of the interferometer mirrors, resulting in phase modulation.

Signal Detection

Sound waves generated directly or indirectly in the gas phase are detected usually by condenser or electret microphones. Detection of sound waves by microphones in condensed matter is typically not suitable. Due to high acoustic impedance mismatches, less than 10^{-4} of the acoustic energy is transferred from a solid sample to the adjacent gas phase. In pulsed excitation of condensed matter, the application of microphones is additionally hampered due to their restricted bandwidth. Therefore, piezoelectric transducers are employed in many cases for the detection of ultrasonic pulses in liquid and solid samples. Quartz crystals, piezoelectric ceramics such as lead zirconate titanate (PZT), lead metaniobate, and lithium niobate as well as piezoelectric polymer films can be applied to the detection of laser-induced shock pulses.

The most common piezoelectric polymer is polyvinylidene fluoride (PVDF), which is available in different thicknesses ranging from 5 to 100 μm as a transparent film or coated with metals for electrical contact. As the sensitivity of piezoelectric detectors usually increases with their thickness, in general, piezoelectric ceramics are more sensitive than thin polymer films. In piezoceramics, eigenmodes are excited by pressure pulses, leading to acoustic waves that decay exponentially within microseconds to milliseconds. Further pressure pulses reaching the detector within this time are overlaid by the signal from the first pulse. This is not observed when nanosecond pressure pulses are detected by thin piezoelectricpolymer films. Therefore, ceramics are preferable for sensitive quantitative analyses in liquids, whereas piezoelectric polymer films are used, if the temporal pressure distribution is of importance, in pulsed PA depth profiling, for example in the analysis of liquid samples by pulsed PAS, in most cases the so-called forward mode is employed for signal detection. In this mode, excitation and detection are performed on different sides of the sample. In weak absorbers, attenuation of the laser beam inside the sample can be neglected. Thus, the laser builds a cylindrical acoustic source resulting in cylindrical waves, which can be detected perpendicularly to the laser beam. In opaque samples, low optical penetration depth leads to a punctiform acoustic source that generates spherical waves. Here, detection both along the laser beam and perpendicular to it is possible. In pulsed PA analysis of solid samples, the generated pressure pulses are often detected in backward mode, where excitation and detection are performed at the same side of the sample. Since piezoelectric transducers are generally not transparent, illumination through the piezo and detection at the same point are not possible. The PA sensor uses a transparent prism as coupling material for both illumination of the sample and transfer of the acoustic energy to the detector. Another possibility is to illuminate the sample by means of an optical fiber and to detect the pressure pulses by a piezoelectric ring around the fiber.

In addition to conventional acoustic detection based on microphones or piezos, optical detection schemes for pressure waves are described in the literature. Optical microphones are applied in PA gas phase analysis and similar fiber optic sensors are reported for liquid phase analysis. Here, an optical fiber is wrapped onto a PA cell. Pressure changes inside the cell cause refractive index changes in the fiber which lead to phase shifts in the light coupled into the fiber and these can be detected, for example, by interferometry. In this case, the fiber is one arm of an interferometer.

In PA investigations of condensed matter, optical methods can be employed for the detection of refractive index changes at an interface of the sample or measurement of the surface displacement caused by a pressure wave. In the first case, the sample is placed at the base of a transparent prism. A continuous probe laser beam is reflected at the interface between prism and sample. A pressure wave transmitted through this interface will change the refractive index, resulting in variations of the optical reflectance, which can be detected optically. If a time-gated video camera is employed for detection, the pressure waves can be imaged in a two-dimensional and time-resolved fashion.

In the second case, the probe beam is one arm of an interferometer. Pressure waves reaching the interface will cause surface displacements in the nanometer range, which can be detected due to changes in the interference pattern. Such optical methods are advantageous due to their ability to perform non-contact measurements and their improved lateral resolution compared to piezoelectric detectors. For three-dimensional PA imaging, two dimensional piezoelectric detector arrays have been proposed, but small piezosensors with fine spacing are hard to realize.

In optical detection, lateral resolution is only limited by the diameter of the probe beam, which can be less than 1 μm. Pressure waves generated in liquid and solid samples can be visualized directly by Schlieren photography or dark field imaging. These techniques allow the influence of optical properties and the illuminated volume fraction of the sample on the produced acoustic waveforms to be investigated. Since the sample needs to transmit light beams in order to be able to visualize the pressure waves, these techniques are restricted to transparent and non-scattering samples. To overcome this limitation, an interesting set-up was proposed: an acoustic lens system was used to image the initial transient pressure distribution inside the sample into a water container, where the three-dimensional pressure distribution could be detected by dark field imaging. In this way, absorbing objects inside a scattering matrix could be visualized as stereo images.

Instrumentation of Photoacoustic Spectroscopy

Radiation Sources

Radiation source can be output from a laser, a monochromator furnishing radiations in UV, IR, or a FT-IR spectrometer (tungsten lamp, carbon arc lamp, high pressure xenon lamp, Nernst glower and lasers.) All radiation must be pulsed at an acoustical frequency 50-1200Hz. PA cell is filled with transparent gas often air or helium and cell volume is kept small, less than 1cm3 in order to preserve the strength of the acoustical signal. In commercial photoacoustic spectrometers, incoherent sources such as lamps are employed in combinations with filters or interferometers. Devices equipped with a small light bulb, with either a chopper or direct current modulation as modulated source and appropriate filters to avoid absorption interferences with other species, are used as compact gas sensors, e.g. for indoor CO_2 monitoring. However, since the generated photoacoustic signal is proportional to the absorbed (and thus to the incident) radiation power, powerful radiation sources, particularly lasers offering high spectral brightness, are advantageous for achieving high detection sensitivity and selectivity in spectroscopic applications. Diode lasers have so far only rarely been employed in photoacoustic spectroscopy owing to their limited power. This situation may, however, change with the ongoing developments in this field. On the one hand, near infrared diode lasers with sufficient power for PAS are available for monitoring overtones and combination bands of molecular fundamental absorptions. On the other hand, current efforts focus on the implementation of widely tunable narrowband all-solid-state laser devices in the mid-IR region for accessing the (much stronger) fundamental absorptions. Optical parametric oscillators (OPOs) and difference frequency generation (DFG) in nonlinear crystals are certainly of great interest for compact spectrometers. Furthermore, recent developments in quantum cascade lasers look very promising in this respect.

Modulation Schemes

Modulation schemes can be classified into the modulation of the incident radiation and modulation of the sample absorption itself. The first technique includes the most widely used amplitude modulation (AM) of continuous radiation by mechanical choppers, electro-optic or acousto-optic modulators as well the modulation of the source emission itself by current modulation or pulsed excitation. In comparison to amplitude modulation (AM), frequency modulation (FM) or wavelength modulation (WM) of the radiation may improve the detection sensitivity by eliminating the continuum background caused by a wavelength – independent absorption, e.g. absorption by cell windows, known as window heating. This type of modulation is obviously most effective for absorbers with narrow line width and most easily performed with radiation sources whose wavelength can rapidly be tuned with a few wave numbers. Pulsed excitation is often applied for liquids but is also of interest for gaseous samples because it permits time gating and the excitation of acoustic resonances.

Photoacoustic Cell

The Photoacoustic cell serves as a container for the sample under study and for the microphone or other device for the detection of the generated acoustic wave. An optimum design of the Photoacoustic cell represents a crucial point when background noise ultimately limits the detection sensitivity. In particular, for trace gas application many cell configurations have been presented including acoustically resonant and non-resonant cells, single and multipass cells, as well as cell placed intracavity. Nonresonant cells of small volume are mostly employed for solids samples with modulated excitation or for liquids and gaseous samples with pulsed laser excitation. As a unique example, a small volume cell equipped with a 'tubular' acoustic sensor consisting of up to 80 signal miniature microphone has been developed. These microphones are arranged in eight linear rows with ten microphones in each row. The row are mounted in a cylindrical geometry parallel to the exciting laser beam axis and located on a circumference around the axis. This configuration is thus ideally adapted to the geometry of the generated acoustic waves.

Detection Sensors

The acoustic disturbances generated in the sample are detected by some kind of pressure sensor. In contact with liquid or solid samples these are piezoelectric devices such as lead zirconate titanate (PZT), $LiNbO_3$ or quartz crystals with a typical responsivity R in the range of up to V bar^{-1} or thin polyvinylidene-difluoride (PVF_2 or PVDF)- foils with lower responsivity. These sensors offer fast response times and are thus ideally adapted for pulsed photoacoustics.

For studies in the gas phase, commercial microphones are employed. These include miniature electrets microphones such as Knowles or Sennheiser models with typical responsivities (R_{mic}) of 10-20 mV Pa^{-1} as well as condenser microphones, e.g. Bruel & Kjaer models with typical R_{mic} of 100 mV Pa^{-1}. Usually R_{mic} depends only weakly on frequency.

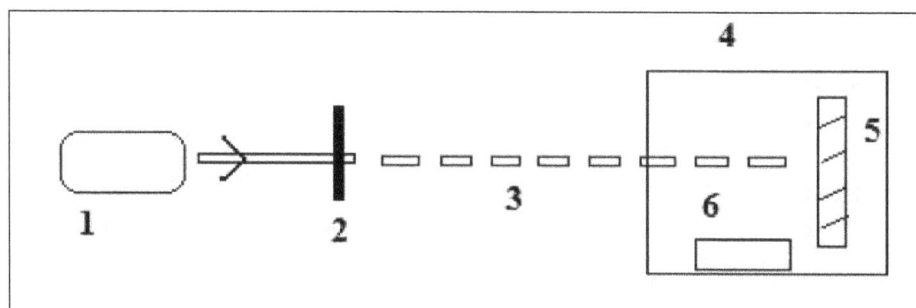

Figure: Typical experimental arrangements used for Photoacoustic spectroscopy. (1) light source (2) the chopper (3) chopped light (4) photoacoustic cell (5) sample (6) Microphone (detection sensors).

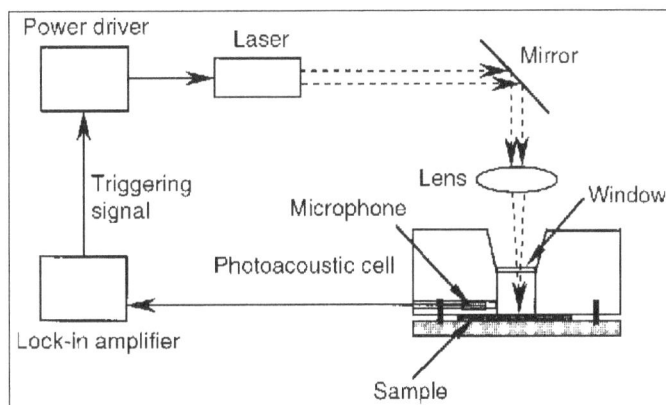

Figure: Photoacoustic effect from a sample.

Experimental Arrangement

As the photoacoustic and related photothermal phenomena comprise a large diversity of facets, there exist a various detection technique which rely on the acoustic or thermal disturbances caused by the absorbed radiation, the selection of the most appropriate scheme for a given application depends on the sample, the sensitivity to be achieved, ease of operation, ruggedness, and any requirement and any require for non-contact detection, e.g. in aggressive media or at a high temperature and/or pressure.

Experimental schemes for photoacoustic studies on solid sample includes the measurement of the generated pressure wave either directly in the sample with a piezoelectric sensor for the pulsed regime, or indirectly in the gas which is in contact with the sample by a microphone.

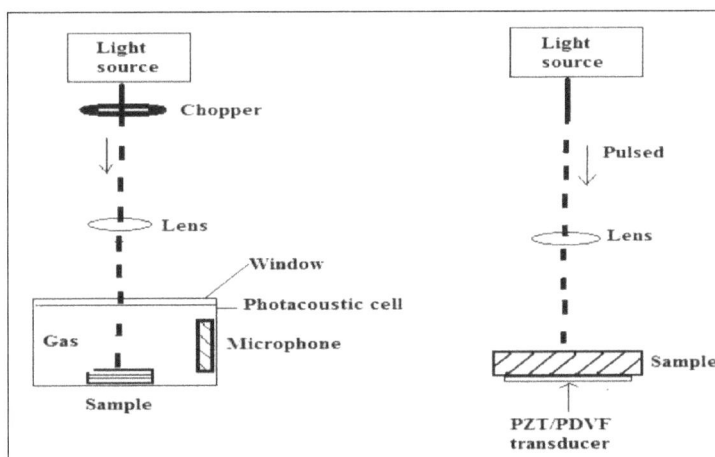

Figure: Typical experimental arrangements used for Photoacoustic studies on solids.

The typical experimental arrangement for the absorption spectroscopy in weakly absorbing liquids is shown schematically in figure. The beam of a pulsed tunable laser is directed through the PA cell that contains the sample under study. The generated acoustic waves are detected by a piezoelectric transducer with fast response time. Usually, only the first peak of the ringing acoustic signal is taken and further processed. Pulse-to-pulse variations of the laser power are accounted for by normalizing the piezoelectric signal with the laser power measured with the power meter after the cell.

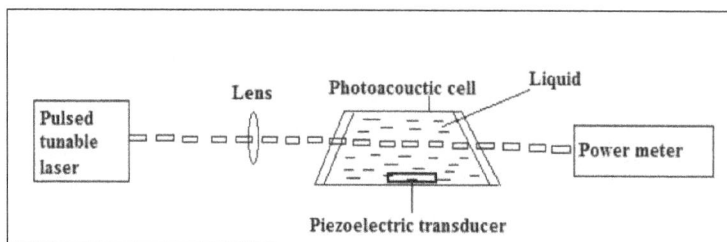

Figure: Typical experimental arrangements used for photoacoustic detection in liquids.
(PZT detection of acoustic wave generated by pulsed radiation in weakly absorbing liquids).

The typical setup for gas phase measurement is shown in Figure below. A tunable laser with narrow line width or a conventional (broadband) radiation source followed by optical filters is used. In general, amplitude-modulated (or sometimes pulsed) radiation is directed through the PA cell. The acoustic sensor is usually a commercial electret microphone or a condenser microphone. These devices are easy to use and sensitive enough for trace gas studies with very low absorption. Often, the detection threshold is neither determined by the microphone responsivity (R_{mic}) itself nor by the electrical noise but rather by other sources (absorption by desorbing molecules from the cell walls, window heating, ambient noise, etc.). However, if this later background is known from reference measurements, the ultimate detection sensitivity is determined solely by fluctuations of the radiation intensity, and by microphone and amplifier noise. The frequency dependence of $Rmic$ is usually rather small and the temperature dependence may have to be taken into account in special cases only. If modulated radiation is employed the microphone signal is fed to a lockin amplifier locked to the modulation frequency. If pulsed radiation is employed, the microphone bandwidth is often not sufficient to resolve the temporal shape of the generated acoustic pulses. However, common microphones can still be used even for nanosecond laser pulses because the length of a single acoustic pulse is essentially determined by the transit time of the acoustic wave across the beam radius.

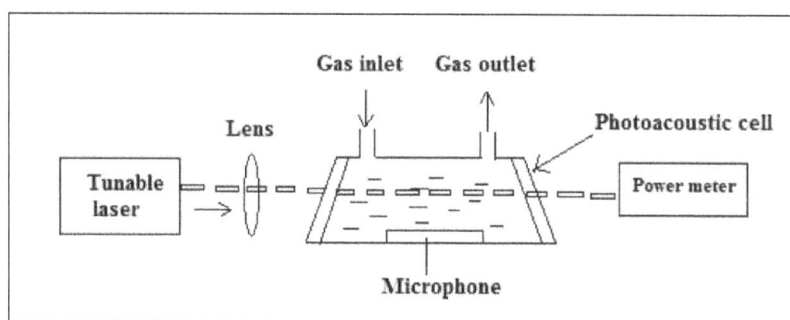

Figure: Typical experimental photoacoustic arrangements for gas monitoring with tunable laser sources.

Applications

1. Depth profiling of mammalian cells for localization of ligands: Phase-resolved monitoring of photoacoustic signals can provide information about the depth profile of a sample. The principle of photoacoustic spectroscopy has been used to determine the depth profiles of ligands and antitumor agents in mammalian cells. Measurements of the in-phase and quadrature components of the photoacoustic spectra (which yield information from the surface and the interior, respectively) of a tumor cell line, AK-5, treated with the antitumor agent coralyne chloride have been made. They clearly show that the drug accumulates in the cell interior and is not seen on the cell surface,

providing in situ evidence for the localization of this drug. Histochemical dyes which stain cells uniformly give identical in-phase and quadrature spectra; spectra of cells incubated with nuclear stains demonstrate a differential staining of the nucleus and the cytoplasm. These results demonstrate the usefulness of phase-resolved photoacoustic spectroscopy in monitoring differential interactions of drugs and other ligands with cells.

2. The biophysics approach: photoacoustic spectroscopy to study the dna-ethidium bromide interaction: The binding processes of ethidium bromide interacting with calf thymus DNA using photoacoustic spectroscopy have been examined. These binding processes are generally investigated by a combination of absorption or fluorescence spectroscopies with hydrodynamic techniques. The employment of photoacoustic spectroscopy for the DNA ethidium bromide system identified two binding manners for the dye. The presence of two isosbestic points (522 and 498 nm) during DNA titration was evidence of these binding modes. Analysis of the photoacoustic amplitude signal data was performed using the McGhee-von Hippel excluded site model. The binding constant obtained was $3.4 \cdot 108$ M(bp)$^{-1}$, and the number of base pairs excluded to anotherdye molecule by each bound dye molecule (n) was 2. A DNA drug dissociation process was applied using sodium dodecyl sulfate to elucidate the existence of a second and weaker bin ding mode. The dissociation constant determined was 0.43 mM, whose inverse value was less than the previously obtained binding constant, demonstrating the existence of the weaker binding mode. The calculated binding constant was adjusted by considering the dissociation constant and its new value was $1.2 \cdot 109$ M(bp)$^{-1}$ and the number of excluded sites was 2.6. Using the photoacoustic technique it is also possible to obtain results regarding the dependence of the quantum yield of the dye on its binding mode. While intercalated between two adjacent base pairs the quantum yield found was 0.87 and when associated with an external site it was 0.04. These results reinforce the presence of these two binding processes and show that photoacoustic spectroscopy is more extensive than commonly applied spectroscopies.

3. Photoacoustic spectroscopy to study the presence of aromatic amino acid in proteins: The examination of aromatic amino acid in six proteins with well known structures using absorption spectra of near ultraviolet PAS over the wavelength range 240–320 nm have been performed. The fundamental understanding of absorption of light and a subsequent release of heat to generate a transient pressure wave was used to test the concept of monitoring aromatic amino acids with this method. Second derivative spectroscopy in the ultraviolet region of proteins was also used to study the regions surrounding the aromatics and the percentage area in each band was related in order to determine the contribution in function of the respective molar extinction coefficients for each residue. Further investigation was conducted into the interaction between sodium dodecyl sulphate (SDS) andbothropstoxin-I (BthTx-I), with the purpose of identifying the aromatics that participate in the interaction. The clear changes in the second derivative and curve-fitting procedures suggest that initial SDS binding to the tryptophan located in the dimer interface and above 10 SDS an increased intensity between 260 and 320 nm, demonstrating that the more widespread tyrosine and phenylalanine residues contribute to the SDS/BthTx-I interactions. These results demonstrate the potential of near UV-PAS for the investigation of membrane proteins/detergent complexes in which light scattering is significant.

4. Photoacoustic Spectroscopy as a key Technique in the Investigation of nanosized Magnetic Particles for Drug Delivery Systems: The study includes how cubic ferrite nanoparticles, suspended as

ionic or biocompatible magnetic fluids, can be used as a platform to built complex nanosized magnetic materials, more specifically magnetic drug delivery systems. The study shows the use of the photoacoustic spectroscopy as an important technique in the investigation of key aspects related to the properties of the hosted nanosized magnetic particle.

5. Analysis of Biological Material: Conventional spectroscopy does not yield satisfactory spectra because of the string light scattering properties of the blood cells, protein and lipid molecules present. PAS permits spectroscopic studies of blood without the necessity of a preliminary separation of these large molecules.

Other biological application of PAS include identification of bacteria states, study of animal and human tissues including teeth, bone, skin, muscle etc., analysis of drug in tissues, investigating of the photo-oxidative decay in human eye lenses etc.

6. Evaluation of Elasticity and Integrity of Pharmaceutical Tablets: A nondestructive method based on pulse photoacoustics was applied for evaluation of elasticity and integrity of pharmaceutical tablets. Variations in porosity, density and sodium chloride content of microcrystalline cellulose tablets were found to be related to parameters extracted from the through-transmitted ultrasonic wave forms. By using the amplitudes and ultrasonic velocities of these wave forms, it was possible to obtain values of a transverse to longitudinal amplitude ratio, and also elastic parameters, such as Young's and shear moduli, for the tablets. The method is a promising tool for evaluating the elastic properties of tableting materials and the structural variations in tablets.

Quantitative analysis of drug content in semisolid formulation using step scan ftir photoacoustic spectroscopy. Step-scan FT-IR photoacoustic spectroscopy in conjunction with a phase modulation technique (modulation frequency of 25 Hz) and digital signal processing was applied in order to quantify the content of brivudin and dithranol in vaseline/drug ointment. The PA spectra of the mixtures exhibit an excellent signal to noise ratio and bands belonging to the drugs are clearly observable down to an 0.5 wt% concentration of the drug. The integrated intensity of the drug bands with reference to a vaseline band was used as a measure of drug content. For comparison, the concentration of drug was determined by capillary zone electrophoresis for brivudin and by HPLC for dithranol. It appears that Beer's law is fulfilled for the PAS data. No sample preparation was required for the PAS experiments.

7. Monitoring Electron Transfer by Photoacoustic Spectroscopy: The electron transfer process between octaethylporphin and quinone molecules dispersed in a polymeric matrix was studied by the photoacoustic technique. It was observed that there was an enhancement of the octaethylporphin photoacoustic signal with an increase of the quinone concentration in the films. This increase appeared to be complementary to octaethylporphin fluorescence quenching and was associated with the electron transfer process. The data were analyzed according to the theory developed by Kaneko for fluorescence data.

8. Gas Phase Analysis: The development of new PA setups for on-line gas monitoring has been achieved through new developments in diode lasers. Atmospheric pollutants that can be detected by PA measurement techniques includes sulfur oxides (such as SO_2), nitrogen oxides (NO_2), carbon oxides (CO and CO_2), hydrogen sulfide, ammonia, methane, and aerosol particles (such as soot).

9. Analysis of Condensed Matter: The main application of UV/Vis.–PAS is the characterization of semi conducting materials. As the PA signal depends on heat diffusion, the thermal diffusivity can also be determined, which is strongly sensitive to the structural quality of the semi conducting material. Furthermore, packaging materials have been characterized by PA measurements in the UV/Vis range. Using depth-resolved PAS, it was possible to estimate the thicknesses and moisture contents of varnish layers on base paper.

Analysis of highly concentrated textile dyes using photoacoustic spectroscopy - The concentrations of textiles dyes are in the range of more than 5 g L^{-1}, resulting in absorption coefficients of 103 cm^{-1}. The combination of extremely high absorption and scattering particles in the dye solution makes a classical transmission spectroscopic analysis impossible. PA spectroscopy is a viable approach to overcome the problems.

10. The Malaria Parasite Monitored by Photoacoustic Spectroscopy: Noninvasive photoacoustic spectroscopy was used to study the malarial parasites Plasmodium chabaudi and Plasmodium berghei, their pigment and ferriprotoporphyrin IX, which is a byproduct of the hemoglobin that the parasite ingest. The result indicate that the pigment consits of ferriprotoporphyrin self aggregates and a noncovalent complex of ferriprotoporphyrin and protein. Spectra of chloroquine-treated parasites reveal in situ interaction between the drug and ferriprotoporphyrin. Chloroquine-resistant parasites, readily distinguishable by this method, appear to degrade hemoglobin only partially.

Laser-Induced Breakdown Spectroscopy

Laser-Induced Breakdown Spectroscopy (LIBS) is an atomic emission spectroscopy technique which uses highly energetic laser pulses to provoke optical sample excitation.

The interaction between focused laser pulses and the sample creates plasma composed of ionized matter. Plasma light emissions can provide "spectral signatures" of chemical composition of many different kinds of materials in solid, liquid, or gas state. LIBS can provide an easy, fast, and in situ chemical analysis with a reasonable precision, detection limits, and cost. Additionally, as there is no need for sample preparation, it could be considered as a "put & play" technique suitable for a wide range of applications.

Considerable progress has been made during the last few years on very different and versatile applications of LIBS, including remote material assessment in nuclear power stations, geological analysis in space exploration, diagnostics of archaeological objects, metal diffusion in solar cells, and so forth. Today, LIBS is considered as an attractive and effective technique when a fast and whole chemical analysis at the atomic level is required.

Plasma Physics of LIBS

Understanding the plasma physics of LIBS is essential to provide an optimized setting for LIBS measurements. A large number of environmental factors affect the plasma life time and features, changing the spectral emission and the performance of this technique for chemical analysis at the atomic level.

Laser Ablation and Plasma Physics of LIBS

Laser-matter interactions are governed by quantum mechanics laws describing how photons area absorbed or emitted by atoms. If an electron absorbs a photon, the electron reaches a higher energy quantum mechanical state. Electrons tend to the lower possible energy levels, and in the decay process the electron emits a photon (deexcitation of the atom). The different energy levels of each kind of atom induces different and concrete photon energies for each kind of atom, with narrowband emissions due to the quantization, with an uncertainty defined by Heisenberg uncertainty principle. These emissions are the spectral emission lines found in LIBS spectra and its features and their associated energy levels are well known for each atom.

If the energy applied to the atom is high enough (overcoming the ionization potential), electrons can be detached by the atom inducing free electrons and positive ions (cations). Initially, the detached electron is the most external one (the furthest with respect to the nucleus) because it has the lowest ionization potential, but with higher energy supply it is possible to detached more electrons overcoming the second ionization potential, the third, and so on. These ions can emit photons in the recombination process (cations absorbs a free electron in a process called free-bound transition) or in the deexcitation process (the cations and the electrons lose energy due to kinetic process in a process called free-free transition). These emissions can be continuum due to the different energies of the ions and the different energy transitions, however cations deexcitation has discrete (or quantized) set of energy levels with characteristic emission lines for each kind of element, allowing its identification together with the atomic emission lines.

The plasma, induced by the interaction pulsed laser-sample, emits light which consists of discrete lines, bands, and an overlying continuum. These discrete lines, which characterize the material, have three main features; wavelength, intensity, and shape. These parameters depend on both the structure of the emitting atoms and their environment. Each kind of atom has some different energy levels which determine the wavelength of the line. Besides the identification of the elements in the sample, the calculation of the amount of each element in the sample from the line intensities is possible taking in account different necessary conditions fixed by local thermodynamic equilibrium (LTE condition) or problems related with matrix effects which can reduce the accuracy of quantitative analysis.

On the other hand, the intensity and shape of the lines depend strongly on the environment of the emitting atom. For not too high plasma densities, both the natural broadening (due to Heisenberg's uncertain principle) and the Doppler broadening (Doppler Broadening is due to the thermal motion of the emitters, the light emitted by each particle can be slightly red- or blue-shifted, and the final effect is a broadening of the line) dominate the linear shape. For high plasma densities, atoms in the plasma are affected by electric fields due to fast moving electrons and slow moving ions, and these electric fields split and shift the atomic energy levels. As a consequence of these perturbations of the levels, the emission lines are broadened and they change their intensity and shape. This effect is known as the Stark effect and it dominates the line shape for dense plasmas. This broadening together with the different parameters of spectral lines (intensities and shapes) and even the continuum radiation features can be useful to determine plasma parameters, such as electron temperature, pressure, and electron density. These parameters are very important to characterize the plasma, giving information about the physical state of it. Moreover the calculation of these parameters is necessary because the set-up has to be tuned to ensure LTE, key condition for an accurate quantitative analysis .

Basically, there are three stages in the plasma life time. The first one is the ignition process. This process includes bond breaking and plasma shielding during the laser pulse, depending on laser type, irradiance, and pulse duration.

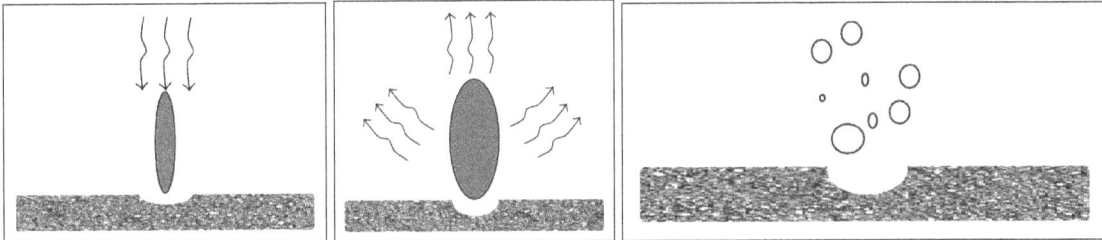

Figure: Plasma life stages: (a) plasma ignition, (b) plasma expansion and cooling, and (c) particle ejection and condensation.

If the selected laser is a femtosecond one, nonthermal processes will dominate the ionization. The pulse is too short to induce thermal effects; hence other effects should ionize the atoms, depending of the kind of sample. The pulse has a huge amount of energy and effects like multiphoton absorption and ionization, tunneling, and avalanche ionization excite the sample. With this amount of energy, the electron-hole created will induce emission of X-rays, hot electrons, and photoemission. This will create highly charged ions through a process called Coulomb explosion. The absence of thermal effects creates a crater with highly defined edges without melted or deposited materials.

In contrast, nanosecond lasers induce other effects. The electron-lattice heating time is around 10^{-12} s, much shorter than the pulse time. This causes thermal effects to dominate the ionization process. Briefly, the laser energy melts and vaporizes the sample, and the temperature increase ionizes the atoms. If the irradiance is high enough, nonthermal effects will be induced too and both will ionize the sample. Between and, plasma becomes opaque for laser radiation, thus the last part of the laser pulse interacts with plasma surface and will be absorbed or reflected, hence it will not ionize much more material. This effect is called plasma shielding and is strongly dependent on environmental conditions (surrounding gases or vacuum) and experimental conditions (laser irradiance and wavelength). This shielding reduces the ablation rate because the radiation does not reach the sample surface. This induces a crater with melted and deposited material around it but at the same time the plasma is reheated and the lifetime and size of plasma is higher.

The next step in plasma life is critical for optimization of LIBS spectral acquisition because the plasma causes atomic emission during the cooling process. After ignition, the plasma will continue expanding and cooling. At the same time, the electron temperature and density will change. This process depends on ablated mass, spot size, energy coupled to the sample, and environmental conditions (state of the sample, pressure, etc.).

If the plasma is induced in a vacuum, the plasma-plume expands adiabatically and the expansion of the ablated material can be described by the Euler equations of hydrodynamics. In contrast, if the surrounding medium is a gas or a liquid, the plume will compress the surrounding medium and produce shockwaves. In this situation the plasma plume is a mixture of atoms and ions from both vaporized material and ambient gas. The shockwave expansion can be described by Sedov's theory.

Plasma temporal evolution changes with pulse duration. For pulses longer than about 5ps, the laser-plasma interactions result in plasma heating and the plasma temperature increases with pulse duration. For short times (below 30 ns) a fs-induced plasma emission intensity decays while ns-induced plasma becomes hotter. The last part of the nanosecond laser pulse is absorbed by the plasma, reheating it, elongating lifetime, and increasing line emission, but at the same time, the background is higher and this decreases the sensitivity for nanosecond set-ups.

A good way to improve the performance is to use a femtosecond laser to ablate the sample and nanosecond or picosecond pulses to reheat the plasma.

The last stage of the plasma life is not interesting for LIBS measurements. A quantity of ablated mass is not excited as vapor or plasma, hence this amount of material is ablated as particles and these particles create condensed vapor, liquid sample ejection, and solid sample exfoliation, which do not emit radiation. Moreover, ablated atoms become cold and create nanoparticles in the recombination process of plasma.

Plasma Emission Spectra

The emission lines from the atomic species can be hidden by continuum radiation that is caused by two processes. The first one is due to radiative recombination. Both continuum and line photons can be produced in such a recombination event as the electron passes from the free state into the upper bound levels of the ion and then cascades down to form a ground state ion. The other effect involved in continuum radiation is called Bremsstrahlung. This effect is related to free-free transitions corresponding to the loss of kinetic energy by an electron in the field of an ion. The electron loses energy in the deceleration process when it travels into the field of the ion, emitting photons in different wavelengths, depending of its initial energy and the loss of this energy. Continuum emission can hide atomic emission peaks, thus this effect should be avoided.

The continuum emission depends on both temperature and density of plasma. These parameters are too high in the initial stages of plasma, especially in the ignition. For this reason, the time control unit has to delay the acquisition window in order to avoid this continuum. For femtosecond lasers, the continuum emission is observed within one nanosecond after the laser ablation; hence the delay with these lasers should be greater than this time. For nanosecond lasers, atomic emission occurs after 1 microsecond and molecular emission occurs at later times from recombination of species in the plasma. Inside this acquisition window, the initial stages of plasma life are characterized by a higher temperature and electron density. These parameters provide a better emission of ionic lines, for that reason these initial stages are better to acquire ionic lines despite the continuum emission. These optimal acquisition windows depend strongly on the sample material and both environmental and experimental conditions, but the values given above can be interesting as a starting point for each kind of laser.

Lifetime for femtosecond-induced plasmas is shorter and with less background. For that reason, the acquisition window size should be shorter for short pulses and femtosecond lasers are better for nongated measurements because the background is weak and the LIBS sensitivity improves.

Generally, femtosecond lasers are better to obtain highly accurate craters and for nongated measurements, and using delays of a few nanoseconds and small acquisition windows can improve LIBS sensitivity. Nanosecond lasers create a melted crater and need delays of a few microseconds and larger windows with the advantages of low complexity in the laser system. There are other kinds of lasers, such as picosecond lasers, for which the time pulse is between a femtosecond and a nanosecond, hence the features are between the two.

LIBS Set-up Components

The main devices involved in a LIBS analysis are shown in. A high-energy pulsed laser (usually in the nanosecond range) is directed at the sample. This light energy vaporizes the sample and induces the plasma.

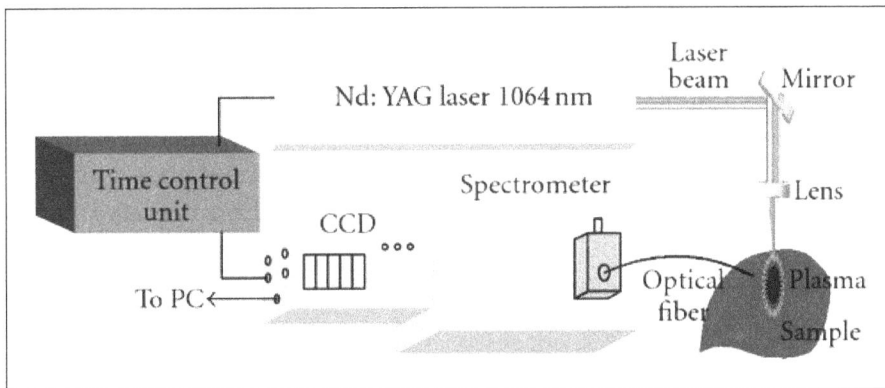

Figure: Typical LIBS set-up.

The spectrometer is in charge of diffracting the light collected, with a more or less complex optical system, in order to obtain the spectral signature. Then, the light is detected by using devices such as a photomultiplier tube (PMT), a photodiode array (PDA), or a charge-coupled device (CCD). Finally, the acquired spectrum is processed by a computer for further analysis. LIBS set-ups need an accurate time control to avoid some plasma life stages and to improve the spectral signature. The choice of the laser combined with the set spectrometer-detector and time control, adapted to environmental conditions, can determine the success or failure of the experiment.

Laser Configurations for LIBS

The main device of LIBS is the laser. It generates the energy to induce the plasma and mainly determines the plasma features. The main parameters related to the laser are the pulse time (explained above), the energy per pulse, the wavelength, and the number of pulses per burst. Obviously, each application works better with a combination of these parameters. Nanosecond-pulsed lasers are the most common for LIBS. Therefore, the most of this section is related to this kind of laser.

Laser Wavelength

The wavelength influence on LIBS can be explained from two points of view; the laser-material interaction (energy absorption) and the plasma development and properties (plasma-material interaction).

When photon energy is higher than bond energy, photon ionization occurs and nonthermal effects are more important. For this reason, the plasma behavior depends on wavelength in nanosecond LIBS set-ups. In the same way, the optical penetration is shorter for UV lasers, providing higher laser energy per volume unit of material. In general, the shorter the laser wavelength, the higher the ablation rate and the lower the elemental fractionation.

The plasma ignition and its properties depend of wavelength. The plasma initiation with nanosecond lasers is provoked by two processes; the first one is inverse Bremsstrahlung by which free electrons gain energy from the laser during collisions among atoms and ions. The second one is photoionization of excited species and excitation of ground atoms with high energies. Laser coupling is better with shorter wavelengths, but at the same time the threshold for plasma formation is higher. This is because inverse Bremsstrahlung is more favorable for IR wavelengths.

In contrast, for short wavelengths (between 266 and 157 nm) the photoionization mechanism is more important. For this reason, the shorter the wavelength in this range, the lower the fluence necessary (energy per unit area) to initiate ablation. In addition, when inverse Bremsstrahlung occurs, part of the nanosecond laser beam reheats the plasma. This increases the plasma lifetime and intensity but also increases the background at the same time. Longer wavelengths increase inverse Bremsstrahlung plasma shielding, but reduce the ablation rate and increase elemental fractionation (elemental fractionation is the redistribution of elements between solid and liquid phases which modifies plasma emission).

The most common laser used in LIBS is pulsed Nd:YAG. This kind of laser provides a compact, reliable, and easy way to produce plasmas in LIBS experiments. The fundamental mode of this laser is at 1064 nm and the pulse width is between 6 and 15 ns. This laser can provide harmonics at 532, 355, and 266 nm, which are less powerful and have shorter time pulses (between 4 and 8 nm). The fundamental and the first harmonic are the most common wavelengths used in LIBS. This harmonics can be useful to work with different wavelengths in the same environmental conditions, because a lot of Nd:YAG lasers can produce all of them. Other kinds of lasers can be used in LIBS, such as CO_2 or excimer lasers to work in far IR or UV ranges, respectively. Lasers based on fiber or dye technology can reduce the pulse width if the user is attempting to work with picosecond or femtosecond pulses.

Laser Energy

The energy parameters related with laser material interaction are fluence (energy per unit area) and irradiance (energy per unit area and time,). Ablation processes (melting, sublimation, erosion, explosion, etc.) have different fluence thresholds. The effect of changes in the laser energy is related to laser wavelength and pulse time. Hence, it is difficult to analyze the energy effect alone. In general, the ablated mass and the ablation rate increase with laser energy.

The typical threshold level for gases is around and for liquids, solids, and aerosols. These values are for guidance and depend strongly on laser pulse time and environmental conditions, reaching up to around for nitrogen at 760 Torr with a Nd:YAG laser, 1064 nm, 7 ps.

Acquisition Time and Delay

The first stages of LIBS-induced plasma are dominated by the continuum emission. The time gate of decay of this continuum radiation change with a wide range of experimental parameters, such as laser wavelength and pulse time, ambient pressure or sample features. Besides, these experimental parameters fixes set the time periods of atomic emission, the most interesting stage of LIBS plasmas.

Plasma evolution using an IR (Nd:YAG) and a UV (excimer) lasers has been analyzed to discover differences induced by laser wavelength. The plasma continuum emission stage was around 400 ns for UV laser and several microseconds for IR laser. Laser wavelength can affect the selection of delay time (gate delay) and integration time (gate window) and these parameters are essential to optimize the signal to background ratio. The analysis of plasma evolution for Zn and Cd in sand has been analyzed in other works, with an optimal gate delay of 0.5 µs and a gate window of 1.5 µs.

This analysis of optimal gate delay and window can be achieved optimizing the signal to background ratio and the repeatability of this parameter. For a Nd:YAG laser at the fundamental wavelength with power density of approximately $2\,GW/cm^2$, the best compromise between lower relative standard deviation (R.S.D.) and higher signal-to-background ratio was found at a delay of approximately 6 µs. The integration time was fixed at 15 µs.

There are different points of view to optimize the gate delay and window, and the big amount of experimental parameters involved in plasma evolution makes it difficult to recommend valid values for these parameters. The selection should be determined case by case in order to achieve a compromise between high-line intensity and low background. Briefly, for Nd:YAG lasers, both times for gate delay and gate window are in the order of microseconds but this values can change if another kind of results are sought.

Sequences of Pulses: Double Pulse LIBS

Proposed twenty-eight years ago to improve the detection limits of some elements, dual pulse LIBS configuration consists on the sequence of two laser pulses temporally spaced in the order of few nanoseconds or microseconds (depending on laser pulse time, the larger the pulse time, and the larger the time delay between pulses). These two pulses can excite the same area and create two temporally spaced plasmas or the second pulse can reheat the plasma induced by the first one.

Using dual pulse LIBS, atomic emission, signal-to-noise ratio, and limit detection can be improved. Conversely, dual pulse LIBS can complicate the LIBS set-up, although the benefits can justify this complication. These improvements provide better detection limits than single pulse configurations due to the induced atomic emission enhancement. Figure below shows the most common dual pulse configurations.

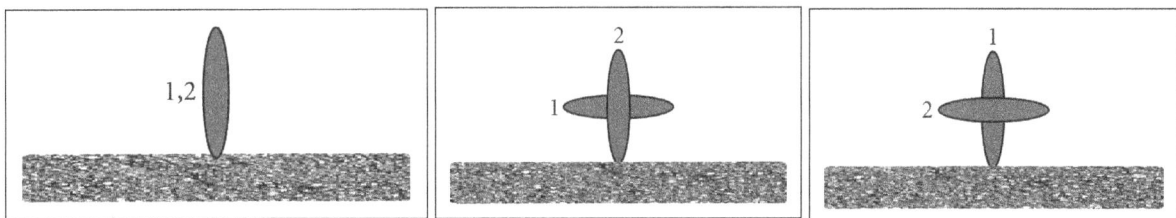

Dual pulse configurations (a) shows collinear configuration; the first pulse ablates the sample and the second one reheats the plasma, (b) is an orthogonal pre-ablative configuration; the first pulse creates a spark on the surrounding media and the second one ablates the sample, (c) shows the same idea as (a), but the plasma is reheated in an orthogonal way.

The signal emission enhancement is due to different factors in these dual pulse configurations. This enhancement in collinear and orthogonal reheating is due to energetic issues. In collinear configurations, the second pulse increases the ablated material and, mainly, reheats the plasma, increasing its volume and its emission, but if the time between laser pulses is enough to allow the fading of the plasma, the second pulse interacts with the sample and ablates more material for the reduced atmosphere induced by the first pulse. In the orthogonal configuration, the second pulse does not ablate more material, but reheats the plasma and reexcites the material ablated by the first pulse. Other authors have given reasons for the signal enhancement on preheating orthogonal configurations, such as a reduction in density or pressure of the surrounding media due to the first pulse.

Spectrometers and Detectors

The spectrometer or spectrograph is a device which diffracts the light emitted by the plasma. There are different designs, such as Littrow, Paschen-Runge, Echelle, and Czerny-Turner. The Czerny-Turner spectrograph is the most common device in LIBS. This spectrograph is composed of an entrance slit, two mirrors, and a diffraction grating. The light comes through the slit and reaches the first mirror which collimates the light, directing it onto the grating. Light is reflected at different angles according to its wavelength. The second mirror focuses the light on the focal plane where the detector is placed.

In recent years, the Echelle spectrograph has been used more extensively. The Echelle spectrograph uses a diffraction grating placed at a high angle, producing a large dispersion in a small wavelength range in each order. As the orders are spatially mixed, a prism is used to separate them. The orders are stacked vertically on the focal plane. For that reason, Echelle devices need a two-dimensional detector. Each vertical portion of the detector contains a part of the spectra and the software composes the whole spectrum.

Different kinds of detectors are used in LIBS, depending of the application. To measure light intensity without spectral decomposition, the photomultiplier tube (PMT) or avalanche photodiode (APD) can be used. On the other hand, for one-dimensional spatial information, the researcher can combine a spectrograph and a photodiode array (PDA) or an intensified photodiode array (IPDA) for time-resolved measurements.

If two-dimensional spatial information is required, the most common devices are charge coupled devices (CCD) and intensified CCD (ICCD). A CCD detector provides less background signal, although ICCD improves the signal-to-noise ratio and is better for time-resolved detection using windows of a few nanoseconds. Another problem related to ICCDs is the price, which is much higher than CCDs.

Collecting and Focusing Light

There are two main parts related to light acquisition and focusing. Typical laser spots currently must be focused to increase the irradiance to induce plasma formation. Light from plasma must then be collected using devices such as optical fibers, lenses, or combinations of both in order to guide it to the spectrometer.

Focusing Laser Pulses with Lenses

As the laser beam width from the majority of solid-state lasers is of the order of 6–8 mm, the most common lenses used to focus laser pulses have diameters of 25 or 50 mm and focal lengths between 50 and 150 mm. There are other applications such as stand-off set-ups where a multilens system or different lenses are required. The material of the lens should have maximum transmission at laser wavelength and they must be coated with antireflection layers to minimize transmission losses (the laser beam can lose around 8% of its energy passing through an uncoated lens. Antireflection coatings can reduce this ratio to around 0.5%).

Collecting Light

A common set-up to acquire plasma light is shown in Figure below. The first lens collimates the emitted light to improve the focalization of the second one into the fiber probe and to optimize the ratio of acquired light. This set-up can be adapted to different systems using only one lens, a multilens device, and even only the fiber probe positioned in front of the spark. The features of lenses used to focus the plasma light are useful for this purpose, adapting them to different environmental features or set-ups.

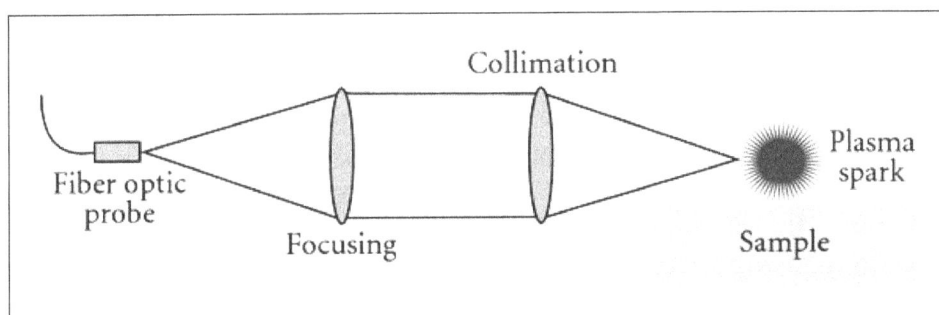

Figure: Typical light acquisition system.

Optical Fibers

Sometimes, the sample is far from the detection system, from the lasers, or from both of them, and so a system based on lenses like the one shown in figure is unpractical. In these situations or when the environment is too aggressive or the access restriction makes it difficult to induce the plasma or acquire the light, fiber optic cables (FOCs) or bundles are used. Although the most common optical fibers in LIBS are made of fused silica (with diameters between 50 μm and 1 mm), different kinds of fibers such as photonic-crystal fibers are used too. Approaches using these coiled fibers (including plastic optical fiber (POF) ones) placed around the plasma plume are also used.

Optical fiber technologies can be used to detect other significant signals from plasma such as shockwaves. Plasma compresses the surrounding media and generates a shockwave. This shockwave can be detected with classical microphones. However, shockwaves can also be detected using other devices based on optical fiber such as Fiber Bragg Grating. In addition, with this technology based on periodical refractive index changes in the fiber core, temperature, or strain, among other factors, can be detected and measured.

Algorithms for Quantitative and Qualitative Analysis

The main goal of LIBS is to achieve a chemical analysis at the atomic level, either qualitatively (i.e., to assess the presence of particular elements) or quantitatively, in which the relative amount of different elements in the sample is quantified from the processing of the acquired spectra. In both cases, a key step is the proper identification of each emission line of a particular element in a neutral or ionized state. If the sample composition is known approximately, the set-up can be adapted to find the optimal spectral range where there are emission lines of the elements under analysis or to discard emission lines of elements outside the sample. At the same time, it is interesting to know the expected relative intensity of each line. If there are doubts about one line that could belong to different elements with very close emission wavelengths, the most probable element can be inferred from the relative intensity and other theoretical and empirical information about emission probabilities. By the same token, the ionization stage is important too. If two elements in different ionization stages are possible for the same emission line, the element in neutral stage should be more likely. Even more, experimental conditions can determine the emission lines in the spectral signature. For example, lines from Fe(I) and Fe(II) are possible in air, but in vacuum, Fe(III) is possible too because the ionization potential of second ionization stage is lower in this medium.

The latest research in LIBS has tried to improve the accuracy of quantitative chemical analysis. Quantitative analysis has to provide the concentration of species in the sample (parts per million), the absolute mass of species (in a particle), or a surface concentration. This analysis usually finishes in a calibration curve, but this curve is strongly dependent on experimental conditions. For this reason, the conditions should be the same when a new sample is analyzed, and sometimes this is difficult because there are many factors that can change them. An averaging of many acquired spectra can attenuate the variation of spectral emission for the analysis, but the pulse-to-pulse variation can be too high to attempt quantitative analysis.

These factors involve aspects both from LIBS set-up and environmental and sample conditions. Laser pulse energy, repetition rate, and detector gain and linearity, as well as lens-to-sample distance (LTSD) can render useless the calibration curve. Calibration curves can be affected by environmental conditions like changes in the atmosphere around the sample or in the light path. At the same time, uniformity in sample composition, surface roughness, or matrix effects affect the accuracy. All these effects induce the variation of the emission intensity between spectra of the same sample.

Matrix effects, that is, the changes in the emission intensity of some elements when the physical properties or composition of the sample varies, greatly affect the accuracy of quantitative estimations in LIBS. For instance, the emission intensity of silicon in water or in soil changes, even if the concentration in both samples is the same. There are two kinds of matrix effects. The first one is called physical matrix effect and is related with the fact that physical properties of the sample can change the ablation parameters. These changes alter the amount of ablated mass; hence the emission intensity changes although the concentration is the same in different matrices. Chemical matrix effects, on the other hand, occur when the emission features of one element are altered by the presence of another one.

One common solution is to use calibration curves, obtained from materials of known composition. A calibration curve is a method for determining the concentration of a substance in an unknown sample by comparing the unknown to a set of standard samples of known concentration. The result is a curve composed by values for the desired quantity as a function of values of sensor output (in LIBS, emission peaks of the element under analysis obtained from LIBS spectra). The curve is created with the measurements of known concentrations and an unknown concentration of a new sample can be predicted using the measurement obtained using its LIBS spectrum. However, many calibrated samples within the expected range of composition variation are required. Another technique, internal standardization, is based on a particular element whose concentration in the sample is constant (it can be added to the sample if needed), and its measurement is used to correct the response to other elements. On the other hand, the impact of other effects (such as unwanted changes in shot-to-shot laser pulse parameters) on the accuracy of the quantitative analysis can be reduced by the so-called external standardization. In this case, an indirect physical magnitude is used to correct (i.e., to normalize) the spectra. For example, it has been observed that during plasma expansion, a shockwave is induced in the surrounding media, and the energy of this shockwave is related to the plasma energy and can be used to reduce the unwanted variations of the spectra. This shockwave can be acquired with a microphone or with other kinds of detectors, such as optical fiber pressure sensors.

Furthermore, there is a technique that solves the problem related to the calibration of LIBS, called calibration-free LIBS. This procedure applies an algorithm in order to avoid matrix effects and the use of calibration curves, providing a fast and accurate quantitative analysis. This algorithm assumes that plasma is LTE, mandatory condition for all kind of quantitative analysis using LIBS, the radiation source is thin and the plasma composition represents the sample composition, typical conditions for LIBS plasmas. The algorithm calculates the line integral intensity between two energy levels of an atomic species, and using different lines at different energy levels, it can calculate the concentration of each element in the sample without calibration from these lines' integral intensities. Theoretically, this algorithm can calculate the concentration of all the elements in the sample up to the detection limit of the method.

Sample Classification Algorithms

The capability of LIBS to detect atomic emission of all chemical elements enables an accurate qualitative chemical analysis. However, there are applications where this kind of analysis is not necessary, for example, if two different materials need to be distinguished and classified. In this case, the raw spectra provided by LIBS can be processed by classification algorithms for distinguishing between different samples.

Algorithms such as artificial neural networks (ANN), support vector machines (SVM), or K-nearest neighbors (Knn) belong to this group. These algorithms need a large training dataset to obtain an accurate classifier and need time to train (SVM and ANN) or need time to classify but not to train (Knn), but the results are usually excellent. These algorithms and other ones can provide a "chemometric" analysis (Powerful statistical signal-processing techniques which provide the automatic identification of chemical information, like spectral fingerprints).

An important aspect of the spectral data provided by LIBS and many other spectroscopic methods is that each spectrum contains a lot of redundant information. In LIBS, in particular, only a few emission lines at particular wavelengths may be interesting to perform the required classification. Redundancy can reduce the accuracy of classifiers and sometimes needs to be avoided. In addition, classification or training time is reduced if the input dataset has less information.

There are two different kinds of algorithms that discard redundant information. The first group is composed of algorithms which select the most discriminating features. These algorithms, such as Sequential Floating Forward Selection (SFFS) or Linear Discriminant Analysis (LDA) select the best features of the input dataset in order to improve the classification ratio (in LIBS, the best wavelengths). The second group includes algorithms which extract features, such as principal component analysis (PCA). These algorithms combine the features of the input dataset and generate new features. Both groups, conveniently adjusted, improve the capability of classifiers.

These algorithms and other ones such as supervised learning algorithms or genetic algorithms have recently been applied to LIBS. These works analyze samples ranging from metallic alloys, soils, heavy metals in water to toxins in toys, and can classify them or detect traces of a particular element, automatically.

A great research effort is being devoted to going beyond a mere spectra-capturing device. With the powerful algorithms described above, an integrated and portable device based on LIBS that can induce the plasma, acquire the spectra, and automatically discriminate or detect materials is feasible. It could result in an essential instrument in many real applications, like the ones that are being explored today.

LIBS Applications

LIBS is useful in a wide range of fields, namely, those which can benefit from a quick chemical analysis at the atomic level, without sample preparation, or even in the field. This paragraph compiles the most important applications at this moment.

LIBS in Archeology and Cultural Heritage

Samples with archeological or cultural value are sometimes difficult to analyze. These samples cannot usually be moved or destroyed for analysis, and some chemical techniques to prepare the sample or a controlled environment in a laboratory are needed. In the first place, portable LIBS devices can be used, solving the problem when the sample cannot be moved. In the second place, LIBS does not need contact to analyze the sample, avoiding damage in valuable samples. Although LIBS ablates an amount of the sample, the crater is nearly microscopic and practically invisible to the eye. In addition, this microscopic ablated surface improves the spatial resolution, providing accurate spatial analysis and even a depth profile analysis of the sample. The sample does not need to be prepared; hence the analysis is clean and fast. Besides, LIBS probes based on optical fibers allow the analysis of samples with difficult access. Despite these facts, LIBS is a microdestructive technique and the researcher should pay attention to experimental parameters in order to avoid critical damage in valuable samples. Many cultural heritage artifacts can be analyzed with the right LIBS set-up.

LIBS is feasible with virtually all types of materials, for instance ceramics, marble, bones, or metals, usually applying quantitative analysis. The most common analysis attempts to determine the elemental composition of the sample in order to help to date it, but it works with bones for analysis of paleodiet. LIBS has been used with delicate samples such as Roman coins or other metallic alloys like bronze, even under water. In the field of painting, it can determine the elements that compose the pigments. This analysis of pigments can help to date and authenticate frescos or paintings. Moreover, LIBS can be used, combined with other techniques, in order to sum the potential of them, such as Raman or X-ray fluorescence (XRF).

LIBS in Biomedical Applications

Biomedicine and LIBS are fields that have not been working together for long. For that reason, this field may provide a large number of new developments in a few years. LIBS can analyze chemical compositions of biological samples such as human bones, tissues, and fluids.

LIBS can help to detect excess or deficiency of minerals in tissue, teeth, nails, or bones, as well as toxic elements. In the same way, cancer detection is possible with LIBS and it can provide a surgical device which can detect and destroy tumor cells at the same time. In addition, classification of pathogenic bacteria or virus is possible too.

The analysis of samples from plants is difficult, because they need a difficult preparation of the sample based on acid digestion processes in order to obtain accurate analysis of micronutrients. LIBS can provide a fast analysis tool with easy sample preparation, for instance in micronutrient analysis of leaves.

LIBS in Industry

LIBS has been targeting many industrial processes for many years, because it is a fast analytical tool well suited to controlling some manufacturing process. Moreover, LIBS can work at a large range of distances, allowing analysis of samples in hazardous and harsh environments. For example, remote detection of explosives has been assessed with LIBS, even at trace levels.

In the nuclear energy industry, the effects of radiation on living beings and devices are widely known. LIBS can work far away from nuclear waste or reactors, using stand-off configuration or with fiber optic probes, avoiding dangerous radiation levels.

In the metallurgical industry, smelters, and final products can reach high temperatures, and LIBS can analyze the alloy compositions in production line or detect impurities in other production sectors, such as the automotive industries.

LIBS can also be useful to detect toxic products like heavy metals in industrial wastes. These waste products should be recycled or stored, and knowing the elements in them can provide key data to reduce the environmental impact of the process.

In the renewable energy field, analysis and detection of impurities in solar cells can be a useful tool to improve the manufacturing processes or to achieve high efficiency solar panels. There are recent research works in this field although there is a huge amount of work to do.

LIBS and Geological Samples: Towards Extraterrestrial Limits

Analysis of some kinds of minerals is possible using LIBS, in particular, of soils and geological samples in situ. Sample features can strongly affect the experimental conditions and reduce the accuracy, but quantitative analysis is still possible. LIBS analysis can detect traces of toxic material in soils, rocks, or water without sample preparation and in the natural environment of the sample.

LIBS can work in a wide range of environmental conditions and with different atmospheres, from air to vacuum. This feature, coupled with the capability to analyze soil samples and the possibility to build a portable set-up, enables the possibility to work in the space. Recently, a spacecraft has been launched to Mars to provide spectral analysis of Mars, geological samples. This spacecraft contains, among other things, a hybrid LIBS-Raman spectrometer.

LIBS Challenges

Probably, the main challenge that LIBS needs to address is its recognition as a standard in chemical quantitative analysis. Calibration-free algorithms offer a good approximation to this goal, but the results are not perfect yet. There are different research lines with the goal of a standard quantitative analysis, attempting to improve the calibration-free algorithm or add new capabilities to it. There are recent works based on spectral normalization to improve the final result or to detect the elements in the sample automatically. This goal may be the most important and could place LIBS definitively among the most widely used spectrochemical techniques.

In order to widen its use in real applications, new advanced and cost-effective instrumentation is required. Currently, a cumbersome and expensive set-up is needed to achieve accurate analysis, and work is in progress to reduce the size and complexity of LIBS set-ups. New work in micro-LIBS and improvements in laser sources useful to LIBS can enable a compact and accurate set-up which makes it feasible in field work. A recent ("hyphenated") approach combines LIBS with other spectrochemical techniques in order to unite the features of them. The Mars Science Laboratory (MSL) is a good example of this because it is a hybrid LIBS-Raman system.

Advances in new techniques and approaches for LIBS analysis, such as optical catapulting and molecular LIBS are being explored. Optical catapulting LIBS (OC-LIBS) uses a pulsed laser below the plasma threshold energy on the sample surface to create a solid aerosol which is analyzed with LIBS. Molecular LIBS, on the other hand, analyzes the emission of molecules resulting from sample ablation or from the recombination between target elements and ambient air. LIBS can improve its performance with this ability and so enable the analysis of organic samples.

Laser Absorption Spectroscopy

There is a range of methods called laser absorption spectroscopy, where laser light is used to precisely measure absorption features of substances. The purpose of such kinds of spectroscopy is frequently to find out details on such substances, but in other cases one utilizes known details of substances for other purposes. For example, laser absorption spectroscopy is often used for realizing optical frequency standards, e.g. by stabilizing the wavelengths of a laser to a precisely defined absorption transition.

Light transmitted through some medium (solid, liquid or gaseous) experiences attenuation due to scattering and absorption; the latter being the dominant loss contribution in many cases. Loss measurements can thus be used to deduct the concentration of a gas of interest with known absorption strength in a gas mixture. If local, in-situ values are desired, the gas sample is sucked into a gas cell and measured there. However, if a satellite measures the sunlight reflected from the earth, this also constitutes absorption spectroscopy, with the atmosphere between the earth and satellite being the gas sampled (remote sensing). The light attenuation for sufficiently small absorption is described by the Beer-Lambert-Law:

$$I = I_0 \cdot e^{-\sigma \cdot N \cdot L}$$

where I is the intensity measured by the detector, I_0 is the original intensity entering the gas cell, σ is the absorption cross section of the absorber (in cm^2 / molec.) and N is the absorber's number density (in molec. / cm^3). Those last two multiplied give the absorption coefficient α ($\alpha = \sigma \cdot N$, in 1 / cm) and L is the path length of the light-matter interaction (also sometimes called "absorption path length").

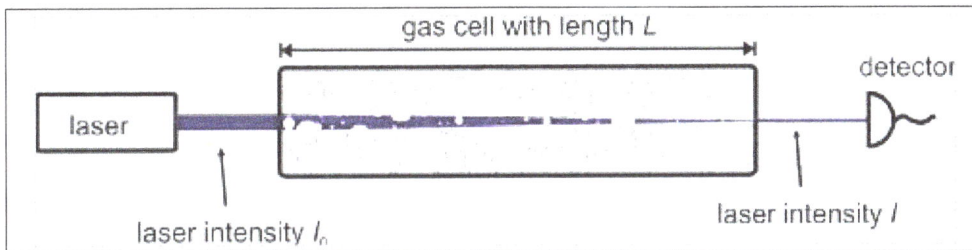

Figure: Simple absorption spectroscopy setup.

When employing absorption spectroscopy it is important to ensure specificity (to be able to relate the measurement signal to only the gas of interest) and sensitivity (to be able to measure small quantities of the gas of interest; these are normally denoted as ppbv (parts-per-billion of volume) or pptv (parts-per-trillion of volume). The latter is often the reason to use optical cavities with high reflective mirrors inside them. These enlarge the path length of the light, increasing the possibility of matter-light interaction and thus the measured signal.

There is a big variety of cavity enhanced absorption methods, called for example ICOS (integrated cavity output spectroscopy), IBBCEAS (incoherent broadband cavity enhanced spectroscopy), OF-CEAS (optical feedback cavity enhanced absorption spectroscopy), TDLAS (tunable diode laser absorption spectroscopy), and also cw-CRDS (continuous wave cavity ring-down spectroscopy). All these methods utilize light reflection by mirrors to enhance the optical path length and increase sensitivity while maintaining a small resonator size.

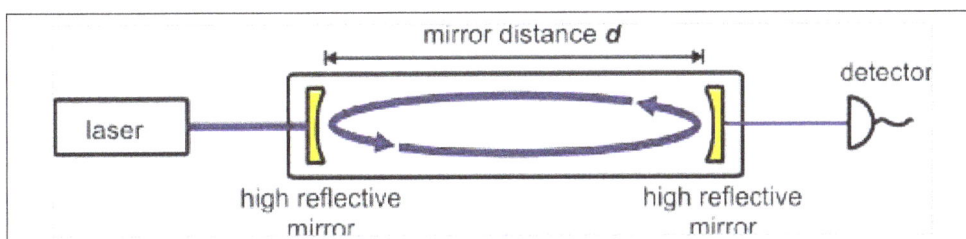

Figure: Enhanced absorption path length by using a cavity with two high reflective mirrors.

There are different methods of calculating the absorption loss from measured signals:

Principles of Laser Absorption Spectroscopy

Direct Absorption Spectroscopy

A frequently used method involves that a tunable narrow-linewidth laser (frequently a single-frequency laser) is tuned through some wavelength range, and the light absorption in some sample is measured as a function of that wavelength. The absorption is often obtained by measuring (a) the optical power of a laser beam which is transmitted through the investigated medium and (b) the optical power of a reference beam (obtained with a beam splitter between the laser and the investigated medium), which is not affected by the medium. That way, one can largely avoid that power fluctuations of the laser (intensity noise) affect the results. In many cases, one uses a balanced photodetector, essentially measuring the difference between two optical powers (rather than their ratio).

Obviously, the obtained spectral resolution is limited by the laser linewidth, which is therefore often minimized with suitable laser designs. Extremely high precision is required in the area of optical frequency metrology, e.g. for realizing extremely precise optical clocks. However, direct absorption measurements are subject to low-frequency laser noise and therefore limited in terms of sensitivity; substantial improvements are possible with modulation spectroscopy.

In some cases, a laser wavelength is swept through a certain range very rapidly ($\rightarrow$ wavelength-swept lasers) for rapid acquisition of absorption spectra. However, the achievable spectral resolution is ultimately limited by the interaction time. Therefore, highly precise measurements often require very slow tuning of the laser.

In some cases, relatively inexpensive tunable diode laser, often in the from of external cavity diode lasers(ECDL), can be used (tunable diode laser absorption spectroscopy, TDLAS). In other cases, substantially more sophisticated laser sources are required, which e.g. can provide far lower line-widths and/or larger tuning ranges.

Setups with Long Propagation Lengths

For highly sensitive detection e.g. of substances in air or other gases, it is necessary to have a substantially long propagation length of light in the gas. Therefore, various types of multipass gas cells (e.g. of Pfund, White or Herriott type) has been developed, where a light beam (typically a laser beam) propagates through the cell many times. That way, long propagation lengths in the gas can be obtained even with a compact optical setup.

Cavity-enhanced Absorption Spectroscopy

Effectively increased propagation lengths can also be achieved by placing a sample inside an optical resonator (cavity). Resonators with high Q factor (high finesse) allow for substantially improved sensitivities.

One possibility is continuous-wave cavity-enhanced absorption spectroscopy, where the optical resonator stays tuned during frequency sweeps of the laser source. The synchronization of laser and resonator may be achieved with an electronic feedback loop or with passive locking techniques, using optical feedback from the resonator to the laser.

Another method is cavity ring-down spectroscopy, where light is injected in the form of a short pulse, and one measures the decay time of light for different cavity lengths, corresponding to different optical frequencies. It is also possible to do such measurements with multiple cavity modes simultaneously.

Intracavity Laser Absorption Spectroscopy

Absorption measurements with particularly high sensitivity can be achieved with intracavity laser absorption spectroscopy, where a sample is placed inside a laser resonator. Typically, one records the laser output spectrum at a certain time after turning the pump source on. The laser would normally exhibit broadband emission with a smooth spectrum, but absorption lines create dips in the spectrum. Due to the large number of resonator roundtrips within the build-up time of radiation in the laser, the method can be very sensitive.

Frequency Modulation Spectroscopy

Absorption features are not always directly investigated by measuring wavelength-dependent absorption, as explained in the section on direct laser absorption spectroscopy. Instead, one may employ frequency modulation spectroscopy (also called wavelength modulation spectroscopy) , where one uses a frequency-modulated laser source. Due to the frequency-dependent absorption, the absorption of the frequency-modulated beam becomes time-dependent, so that a power modulation of the transmitted beam can be detected e.g. with a photodiode. Typically, one does not simply measure the magnitude of the induced power modulation, but rather processes the photodetector signal in an electronic mixer together with the modulation signal.

The optical frequency modulation is in some cases directly obtained by modulating the laser; for example, a laser diode exhibits a frequency modulation (together with an amplitude modulation) when its drive current is modulated. In other cases, one uses an electro-optic modulator as a phase modulator in conjunction with a continuously emitting laser (often a single-frequency laser).

Frequency modulation spectroscopy generally allows for substantially higher sensitivities than direct absorption spectroscopy, essentially because the laser noise limiting the sensitivity is noise around the modulation frequency, rather than around zero frequency. Note that laser noise is usually strong at low frequencies but decays strongly at higher noise frequencies.

Frequency modulation spectroscopy can also be combined with cavity-enhanced methods. For example, for ultra-sensitive measurements one may use an enhancement cavity and a modulation frequency which matches the free spectral range of that cavity.

Two-photon Absorption Spectroscopy

There are cases where laser absorption spectroscopy is not utilizing ordinary linear absorption, but instead two-photon absorption – a process where two photons are simultaneously absorbed, exciting a single atom or ion. The following aspects are relevant for understanding the benefits of that technique:

- In some cases, it is beneficial that optical transitions with relatively high energy differences can be probed without realizing ultraviolet lasers emitting a very short wavelengths.

- One may investigate transitions which are not possible with electric dipole interactions. This is typical for extremely narrow forbidden transitions, which are particularly interesting for precision measurements.

- Two-photon absorption can be Doppler-free, i.e., not affected by the random movement of atoms or ions in a gas cell, if the absorption process involves two counter-propagating photons. The Doppler frequency shifts for both photons are then opposite to each other, so that they cancel out.

Such techniques have e.g. been applied for measuring the frequency of the 1S–2S transition in atomic hydrogen with an extremely high precision.

Absorption Spectroscopy on Molecular Gases

Various aspects of absorption spectroscopy on gases have already been covered above, but some additional remarks are appropriate.

When laser absorption spectroscopy is done on molecular gases, one can identify different species through their different absorption lines (molecular finger prints). This can be used e.g. for detecting trace gases and measuring their concentrations in the atmosphere.

Spectral Regions

One has to decide on the used spectral region:

- The highest sensitivities are possible when using mid-infrared laser sources, because the strongest absorption lines are usually in that spectral region. The optical frequencies directly correspond to vibration or rotation modes of molecules. Unfortunately, mid-infrared sources are relatively difficult to make, thus expensive and limited in performance.

- Alternatively, one may work in the near-infrared spectral region, where laser sources are substantially less expensive and often also exhibit higher performance. In that case, one has to use overtone transitions, with optical frequencies which are integer multiples of molecular vibration frequencies. As these transitions are much weaker, the achievable sensitivity is lower.

Doppler-free Spectroscopy

A fundamental challenge in the spectroscopy of gases is the thermal movement of the gas molecules, which causes Doppler broadening of the optical transitions – often going far beyond the natural linewidth. However, there are various techniques of Doppler-free laser spectroscopy, where such effects are eliminated.

If gas atoms (or molecules) are exposed to a narrow-band laser beam which is tuned to an absorption transition, only those atoms will be excited for which the Doppler shift makes them resonant with the laser beam. For example, if the optical frequency of the pump laser is on the lower end of the Doppler-broadened absorption line, essentially only those atoms will interact with it which are moving towards the pump light with an appropriate velocity. Similarly, a counter-propagating probe beam with the same optical frequency can interact only with atoms moving in the opposite direction. Therefore, the pump beam has virtually no influence on the absorption probed with the probe beam. This changes, however, when the frequency of both beams is tuned to the line center, so that they interact with the same atoms: the absorption for the probe beam is then somewhat saturated (reduced) if the pump beam is sufficiently intense. Therefore, one can detect a narrow dip in the middle of the absorption line as recorded with the appropriate beam. The width of that dip is determined by the natural linewidth, which can be much smaller than the Doppler-broadened linewidth.

Another possibility is Doppler-free Fourier transform spectroscopy involving two frequency combswith slightly different comb line spacings.

Applications of Laser Absorption Spectroscopy

Methods of laser absorption spectroscopy are often used for detecting the composition of materials, often including quantitative measurements of concentrations. Some examples:

- Concentrations of trace gases in the atmosphere are measured e.g. with laser radar (LIDAR) methods in the context of environmental monitoring. Similarly, pollutants can be detected in water, and concentrations of medically active substances can be measured. The Beer–Lambert law can be used to related measured absorption coefficients to concentrations of substances.

- Biology, medicine and chemistry can profit from precise and rapid methods of material analysis, which are partly non-destructive. For example, by measuring concentrations of various substances in human breath, one may retrieve vital information on medical conditions.

- Combustion processes, for example in combustion engines, can be monitored with laser spectroscopy. One may learn about the chemical composition and the gas temperatures, both with sufficiently high temporal resolution.

- Security applications include the detection of explosives and drugs.

Intracavity Laser Absorption Spectroscopy (ILAS)

Intracavity laser absorption spectroscopy is a special laser spectroscopy technique for highly sensitive spectroscopic measurements. The basic principle is as follows. The substance to be evaluated (e.g. some gas sample) is placed within the resonator of a laser, which is preferably based on a gain medium with broad gain bandwidth and a resonator with low losses. When the laser is turned on, it starts to oscillate on many resonator modes simultaneously; only after many resonator round trips will the optical spectrum of the generated light strongly concentrate to the spectral region with highest gain. During this evolution, weak absorption features of the tested sample can imprint signatures on the spectrum, because they can influence the spectrum during many round trips. A measurement of the spectrum is done some time after switching on the laser; this time should be long enough to allow for strong spectral features to develop, but also short enough to prevent too strong narrowing of the spectrum caused by the finite gain bandwidth.

With a carefully optimized setup, intracavity absorption spectroscopy allows for extremely long effective path lengths of tens of thousands of kilometers, and at the same time very high spectral resolution. Suitable laser gain media for the spectroscopy of gases include neodymium-doped fibers and bulk glasses, titanium-doped sapphire, laser dyes, color center crystals, laser diodes, and vertical external cavity surface-emitting lasers. Important issues are to have a broad gain bandwidth, to minimize resonator losses, and to avoid any parasitic reflections within the laser resonator.

Principle of ILAS

The basic principle is as follows: The substance to be evaluated (e.g. some gas sample) is placed within the Resonator of a laser, which is preferably based on a Gain medium with broad gain Bandwidth and a resonator with low losses. When the laser is turned on, it starts to oscillate on many resonator Modes simultaneously; only after many resonator round trips the spectrum of the generated Light will strongly concentrate to the spectral region with highest gain. During this evolution, weak absorption features of the tested sample can imprint signatures on the spectrum, because they can influence the spectrum during many round trips. A measurement of the spectrum is done some time after switching on the laser; this time should be long enough to allow for strong spectral features to develop, but also short enough to prevent too strong narrowing of the spectrum caused by the finite gain bandwidth.

References

- Laser-Spectroscopy: thefreedictionary.com, Retrieved 2 March, 2019

- Laser-spectroscopy: sciencedirect.com, Retrieved 11 May, 2019

- Laser-absorption-spectroscopy: rp-photonics.com, Retrieved 17 January, 2019

- Laserabsorptionspectroscopy: iup.uni-bremen.de, Retrieved 29 July, 2019

- Laser-absorption-spectroscopy: rp-photonics.com, Retrieved 9 February, 2019

- Intracavity-laser-absorption-spectroscopy: rp-photonics.com, Retrieved 23 June, 2019

- Intracavity-Laser-Absorption-Spectroscopy: timbercon.com, Retrieved 29 April, 2019

Permissions

We would like to thank the editorial team for lending their expertise to make the book truly unique. They have played a crucial role in the development of this book. Without their invaluable contributions this book wouldn't have been possible. They have made vital efforts to compile up to date information on the varied aspects of this subject to make this book a valuable addition to the collection of many professionals and students.

This book was conceptualized with the vision of imparting up-to-date and integrated information in this field. To ensure the same, a matchless editorial board was set up. Every individual on the board went through rigorous rounds of assessment to prove their worth. After which they invested a large part of their time researching and compiling the most relevant data for our readers.

The editorial board has been involved in producing this book since its inception. They have spent rigorous hours researching and exploring the diverse topics which have resulted in the successful publishing of this book. They have passed on their knowledge of decades through this book. To expedite this challenging task, the publisher supported the team at every step. A small team of assistant editors was also appointed to further simplify the editing procedure and attain best results for the readers.

Apart from the editorial board, the designing team has also invested a significant amount of their time in understanding the subject and creating the most relevant covers. They scrutinized every image to scout for the most suitable representation of the subject and create an appropriate cover for the book.

The publishing team has been an ardent support to the editorial, designing and production team. Their endless efforts to recruit the best for this project, has resulted in the accomplishment of this book. They are a veteran in the field of academics and their pool of knowledge is as vast as their experience in printing. Their expertise and guidance has proved useful at every step. Their uncompromising quality standards have made this book an exceptional effort. Their encouragement from time to time has been an inspiration for everyone.

The publisher and the editorial board hope that this book will prove to be a valuable piece of knowledge for students, practitioners and scholars across the globe.

Index

www.ingramcontent.com/pod-product-compliance
Lightning Source LLC
Chambersburg PA
CBHW082049190326
41458CB00010B/3488